J. Crockett, C. Harper, A. O'Connor
Series Editor: M. Cay

Starting Home Economics Level 2

Macmillan Education

First published 1985
Reprinted 1986, 1987

Published by
MACMILLAN EDUCATION LTD
Houndmills, Basingstoke, Hampshire RG21 2XS
and London
Companies and representatives
throughout the world

Printed in Hong Kong

British Library Cataloguing in Publication Data
Starting home economics
Bk. 2
1. Home economics
I. Cay, Marjorie
640   TX167
ISBN 0-333-36255-1

Contents

1 What is 'Starting Home Economics Level 2' about?   1

Part 1 Food Studies   1
2 Plant foods   1
3 Fruit and vegetables   3
4 Cereals   13
5 Sugar   22
6 Fat   26
7 Flour mixtures   34

Part 2 Home Studies   53
8 Introduction: how do we furnish our homes?   53
9 How many kinds of materials are there in your home?   57
10 Why do we choose different materials?   61
11 How can we try to keep things looking as good as new?   67

Part 3 Reference Section   78
Microscopes   78
Food labelling laws   83
Food tables   86
Index   89

Acknowledgements

The publishers and authors wish to thank the following photographic sources:

John Darbey and Richard and Sally Greenhill.
Ron Chapman (p69 Left)

Designed and illustrated by Illustra Design.
Additional illustrations by Ursula Sieger.

Cover design: KAG Design.
Cover photograph: Contrast Photographers.

The authors acknowledge the inspiration provided by Rene Finch, and also the help and cooperation provided by colleagues and pupils in their respective schools.

1 What is 'Starting Home Economics Level 2' about?

Can you answer these questions?

1 Why do we need to cook tiny pieces of rice almost as long as a potato which is ten times the size?
2 Is there a semolina plant?
3 If you were a castaway on a desert island, what would you use as a cup?

The answers to these and other vital questions will be found by following Level 2 of Starting Home Economics.
  We will be investigating and using foods from plants, and also materials used in the home in order that you will be able to choose them and use them in the way you like.

Part 1 Food Studies

2 Plant foods

Find out . . .

. . . if you can name the different parts of a plant.

● Look at these drawings and name as many parts of each plant as you can.

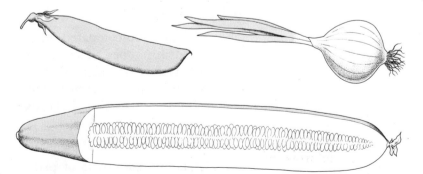

Fig. 1

● If possible look at a complete real plant and name all its parts.
● Use the following information to see if you were right.

1

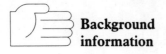

Background information

Parts of plants.

Roots: attach plants to the earth. They pass water and minerals from the soil to the plant. Some roots are swollen.

Bulbs and *tubers*: these are food stores, usually underground, which store food for the new growing plant.

Stems: hold up the plant. They have to be strong but some are very woody.

Green leaves: make food for the plant. They are very watery.

Seeds: contain the new plant or embryo surrounded by a food store. They must contain all the food the new plant needs until it can get its own food.

Fruits: contain seeds. The seeds are usually released when the fruit is ripe.

Flowers: Seeds are produced from the flower.

When you have finished reading this information *close* your book and list three points which you remember. Now compare your list with a partner. Decide on the TWO most important points and report back to your class.

Extra work

Match the vegetables listed below with the part of the plant. The first one has been done for you: the answer is (a). Record your answers in your book in the form of a table.

PARTS OF A PLANT	VEGETABLES
1 leaf	(a) cabbage
2 bud	(b) runner beans
3 root	(c) radish
4 flower bud	(d) celery
5 stem	(e) cauliflower
6 seeds	(f) peas
7 ripe fruit	(g) tomatoes
8 unripe fruit	(h) brussels sprouts.

When you compared the parts of plants you were doing what a botanist does. The names you gave to the parts were the **BOTANICAL** names. When we talk about fruits and vegetables we have to say clearly whether we are using HOME ECONOMICS names or BOTANICAL names.

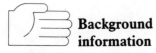

Background information

Classifying fruits and vegetables.

Botanists sort parts of plants into sets according to how they grow. A fruit always contains seeds according to a botanist. It follows that anything which contains seeds is a fruit. So A TOMATO IS A FRUIT.

Home Economists sort plant foods into sets which describe how we eat them. Plants we eat with savoury or salted foods we usually call vegetables, or 'veg.' for short. So, a TOMATO CAN ALSO BE A VEGETABLE.

3 Fruit and vegetables

Vegetables: how do we prepare them?

... the important things you should look for before starting to prepare vegetables. Your teacher will have a range of vegetables such as carrots, parsnips and potatoes for you to look at.

Find out . . .

● Work in two's ● Share all the vegetables out amongst the class, making sure that partners have the same vegetables.
● One partner must cut their vegetable LENGTHWISE.
● The other partner must cut their vegetable CROSSWISE.
● Examine carefully and record everything you see, feel and smell ● Work out how you think the vegetable should be prepared ● Make a report, using drawings if you wish, for the rest of your class ● List the important points you should think about when preparing vegetables. Here are some clues:

(a) Small pieces of vegetable cook more quickly than big pieces.
(b) Some vegetables have hard parts in the centre called cores.
(c) Some vegetables have unpleasant skins.
(d) Some vegetables fall to pieces if cut!

● Discuss your list with the rest of the class. Which points did you all agree on?

Extra things to find out
● Examine large equal sized pieces of potato, carrot, and parsnip, that have been boiled for 5 minutes. Try to answer these questions:

1 Which has cooked the most? What could you have done to the other two varieties to make them cook more quickly?
2 Compare the carrot and parsnip. How are they similar? How are they different? What difficulty could arise if parsnips were cooked whole? How could you solve it?

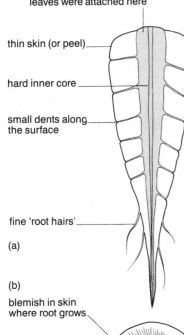

leaves were attached here

thin skin (or peel)

hard inner core

small dents along the surface

fine 'root hairs'

(a)

(b)

blemish in skin where root grows

harder inner core

softer 'outer' area

Fig. 2 (a) Parsnip cut lengthwise; (b) cut crosswise.

Extra work

What advice would you and your partner give to solve these problems? How to:

(a) make finely chopped onions for a salad?
(b) cut onion rings?
(c) take out a core from an apple so it is still whole?

3

(d) slice a tomato so that the seeds do not fall out?

(e) cut a piece of cucumber so there are no seeds?

(f) cut cabbage for a coleslaw?

● Look in recipe books and find out if they give the same solutions.

Practical activity

Making soup.

You have just been cutting vegetables such as potatoes, parsnips and carrots. Use these vegetables to make a quick soup. You can add herbs or a stock cube (e.g. Oxo) or just seasoning to add extra flavour.

Your teacher will tell you how many vegetables to prepare and how much water to add, but look up soups in recipe books to get ideas. For instance you could make a celery soup from:

2 sticks of celery; 1 slice of onion; 250ml water; 1 stock cube; salt and pepper.

Clean the vegetables really well because you do not want to eat dirt. You do not need to peel all vegetables but you do have to take care to inspect each leaf or seed. Look for insects and slugs as well as loose soil.

Your soup is going to be thickened by the vegetables, so they will need to be crushed when they are tender. Think how you might make your soup and then make a flow chart and show it to your teacher. The diagrams shown in Fig. 3 may help you.

Fig. 3 Stages in making soup.

Choose other foods to eat with your soup to make it into a square meal. You could sprinkle grated cheese on top and make some toast.

● Record what you did and say how successful you were at making a quick soup ● Could you make it quicker next time? ● Could you invent other soups?

Find out ...   ... what you do with floppy or slippery vegetables.

Which of the methods shown in Fig. 4 do you think could be successful with an onion?

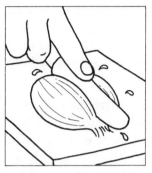 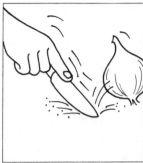  

Fig. 4

Most people have their favourite knives at home, but it is worth while thinking about which knife is best for the job. Of course the knife will not cut properly if you do not hold it properly. But even with the right knife held in the right way, a slippery food will slip on a slippery surface.

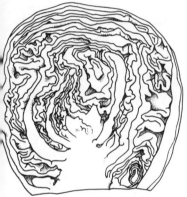
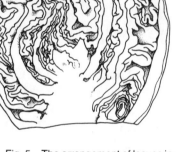

Fig. 5   The arrangement of leaves in a cabbage.

Discussion: cutting up vegetables.

1 Decide the main points to think about before starting to cut up vegetables like onions.
2 Think about how they are constructed, why they break apart, and why they make you cry.
3 Discuss the best knife and the best cutting surface for SAFETY, EFFICIENCY and EASE OF USE.

You may find it helps you to 'see' what could happen if you make yourself a collection of paper onion 'leaves' and wrap one over the other until you have an 'onion'. Other vegetables where one leaf grows over another include cabbages.

Practical activity

● Cut an onion into small, even pieces ● Work safely and efficiently. The diagrams in Fig. 6 will help to remind you of the points you have discussed.

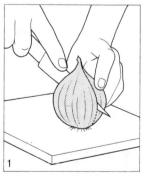 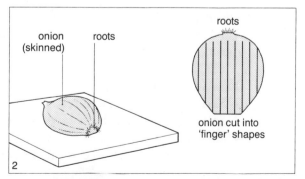 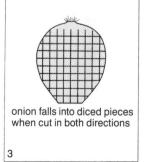

onion
(skinned)   roots

roots

onion cut into
'finger' shapes

roots

onion falls into diced pieces
when cut in both directions

Fig. 6   The three stages of cutting an onion.

Discussion

How would you cut a cabbage into even sized pieces?

5

1 Use your onion and cabbage to make a coleslaw. You can add other vegetables to give more colour and flavour. Make a quick dressing. A coleslaw is nice with an acid dressing, for example, orange juice and a little vegetable oil or lemon juice and/or vinegar and oil. Some people like a natural yoghurt dressing. Whichever you choose, put the vegetables straight into the dressing. Enjoy your salad. What will you eat with it? Will it be a square meal?

2 Look at the salads you can buy from the refrigerated cabinets in a supermarket. Note which foods have been mixed together, and which kind of dressing has been used. Note also that some salads have special names, like coleslaw which is Dutch for cabbage salad. Today we all know that a coleslaw has shredded cabbage in it, but someone had to think of doing it, and then giving the cabbage mixture a name.

Make a list of salads with special names and note the foods which they contain.

What is a salad?

If you look up 'salad' in the dictionary you will find that a salad is:

"A COLD DISH OF VARIOUS MIXTURES OF RAW OR COOKED VEGETABLES OR HERBS USUALLY SEASONED WITH OIL, VINEGAR ETC., AND EATEN WITH OR INCLUDING COLD MEAT, FISH, HARD BOILED EGGS, ETC". (*Concise Oxford Dictionary*).

It takes a lot of words to explain the meaning of the one word 'salad'. That is partly because different people eat different kinds of salad. We also eat salads in different ways – salads can be almost a whole meal in themselves or just a little 'titbit' to eat with other foods.

Discussion: salads.

Take each point of the dictionary definition and discuss what it means.

1 Salads are always cold.
2 Salads are always mixtures.
3 Salads can contain both raw and cooked vegetables.
4 Salads are usually seasoned with oil or vinegar.
5 Salads can include cold meat, fish, hard boiled eggs, etc.

Ask people at home what they mean by the word 'salad'. Discuss how words like 'salad' change their meaning over the years. The dictionary definition given above was taken from a 1976 edition. Look back to your list of salads with special names. Can you find a food in one of them which is not mentioned in the dictionary definition? Which of these special salads contains Mains and/or Fillers?

Practical activity

Make a mixed salad which contains both fruit and vegetables.

• If you want to add crispness you can add chopped apple • If you want to add sharpness you can add orange or grapefruit segments • If you want to add colour and sweetness you can add raisins or sultanas • Make your own selection and show it to your teacher giving reasons for your choice • Say if you would need to eat your salad with another food to get a square meal • Give your salad a name. If it contains your friends' favourite foods it could be 'YOUR GROUP'S' salad.

Discussion
Which foods do we expect to eat together? One class made a note of what people said:

1 'I wouldn't like to eat sweet things with meat'.
2 'It wouldn't be a salad without beetroot'.
3 'I wouldn't like oil on my lettuce leaves'.
4 'We always have salad cream from a bottle'.
5 'In France we have a green salad every day. We eat it after the meat'.
6 'We serve side dishes of things like tomato and cucumber with curry'.
7 'We always have apple sauce with pork'.
8 'We like gooseberry sauce with mackerel'.
9 'I wouldn't like to eat cold food at the same time as hot'.

Salads have become popular because people like their crispness, freshness and colour. But they are also eaten because they help to keep us healthy. To understand why this is so you must learn more about fruit and vegetables.

What is in a can of fruit or vegetables?

Find out ...

... what is in a can of fruit.

• Drain the juice from a can of mandarins, strawberries or grapefruit • Weigh the fruit • Liquidise the fruit and then drain this juice through a sieve • Measure the juice from the sieve • Weigh the solids in the sieve.

• Record what you did in your books • Record all your

7

measurements • Try to account for your results by answering these questions:

(a) Why is the weight of the segments of fruit less than the weight on the can?
(b) Why do the segments weigh less when they have been liquidised?
(c) Where has the liquid collected from the sieve come from?

Practical activity

Make a fruit juice cocktail.

• Mix your juice with other available juices to make a 'cocktail' • Think about the final colour and flavour • Do you want to serve it chilled?

Find out ...

... how the water is hidden inside fruit and vegetables.

You have seen the juice from fruit. It is more difficult to get the juice from some vegetables, but you can see it easily in, say, tomatoes. Scientists see much more using a microscope than we can see with the naked eye. Microscopes make small things big enough to see. So when you look at pieces of plants magnified you can see things like these:

Fig. 7 Microscopic views of (a) raw onion cells; (b) cooked pear cells; (c) cooked garden pea cells.

cells (all glued together in the raw onion)

(a)

veins cut across

cell

separated cell

(b)

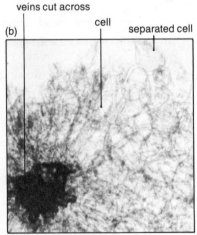

cells (have separated in the cooked pea)

some lose cell contents

(c)

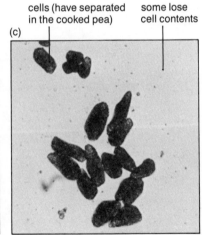

Background information

All plants have cells. You can see them for yourself if you learn to use a microscope. Hopefully by now you have practised making slides and using a microscope, using the information in the Reference Section. Now use your skills to look at the cells of an onion.

Record and label a few cells. Remember your microscope may magnify a different amount. All plant cells have WALLS and JUICE or SAP inside the walls. Make sure you have labelled the cells you have drawn with their cell walls and sap.

Question: Where is the water in a lettuce?

8

Which are the watery foods?

Water is very important to us. Seventy per cent of our bodies is actually water, yet we look pretty solid. We lose about three pints of water each day in urine and sweat. We would die from lack of water before lack of food. We could not last for more than three days without water. When we are ill with a high temperature we need to drink a lot. In our normal diet the watery foods are the ones we can eat plenty of without getting fat.

 Find out . . .

. . . how to find the watery foods in food tables.

We can eat plenty of watery foods without getting fat because water is not fattening. Scientists have measured how much water is in different foods. All of us can look up this information in food tables. Your teacher will show you the tables in books in your Home Economics library. You will need to practise using them, so start with this short table. It tells you how much water is in 100g of each food.

AMOUNT OF WATER IN 100g OF EDIBLE FOOD

MAINS	g	FILLERS	g
Eggs	75	Baked Potatoes	58
Cottage Cheese	79	Mashed Potatoes	77
Edam Cheese	44	Boiled Rice	70
Cheddar Cheese	37	Wholemeal Bread	40
Grilled Back Bacon	36	Toast	24
Cooked Black Pudding	44	Digestive Biscuits	5
Cooked Beef Burger	53	Chocolate Bar	2
		Crisps	3
		Butter	15
		Lard	1
		Sugar	0

F. & V.	g	WATERY DRINKS	g
Flesh of Apple	84	(1ml of watery liquids e.g. water, weighs 1g)	
Banana	71		
Grape	79	Cows Milk (Doubles)	88
Orange	86	Orange Juice	88
Plum	79	Coca-Cola	90
Raw Cabbage	90	Yellow Custard	75
Cooked Peas	80	Fruit Jelly	84
Tomato	93	Vegetable Soup	86
		Salad Cream	53

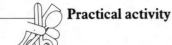 **Practical activity**

Using a water-content food table.

● See if you can use the water-content table to list these foods in order, starting with the most watery.

Apple; Bacon; Lard; Milk; Banana.

In the food table the foods were arranged in groups, but in the list they are jumbled ● When you have found the figure for the water in each food draw a bar graph like this one for digestive biscuits.

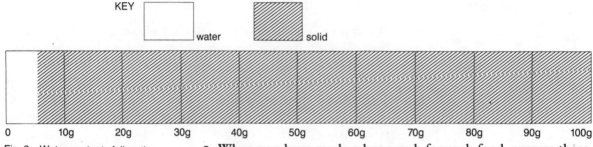

Fig. 8   Water-content of digestive biscuits.

● When you have made a bar graph for each food, answer these questions:

(a) Which food has the most water per 100g?
(b) Which food has the least water per 100g?

Why do we take the water away from foods?

You know now that fruits are watery foods. Have you eaten fruit which has had its water taken away? Have you eaten any other food which has had its water taken away, or which has lost its water naturally? We call all these foods **DRY FOODS**.

 Find out . . .

. . . about how drying changes foods.

● Work in two's to examine these pairs of foods:

(a) dried onion and fresh onion;
(b) dried apple and fresh apple;
(c) stale bread and fresh bread;
(d) sultana and grape.

● Taste them as well as looking at and feeling them. Discuss what you find and try to account for any differences. Note any general differences which were the same for each pair.

Questions
1 Which of each pair had the stronger flavour?
2 Which of each pair was the crisper or tougher?
3 Which was the sweeter?
4 Would you use the same weight of dried as of fresh onion to flavour a soup?

10

5 After they have been dried, what are the following foods called?

(a) garden peas; (b) grapes; (c) plums?

6 (a) Look up the water content of fresh garden peas (cooked), grapes and plums in the table on page 9.

(b) Look up the water content of dried peas, grapes and plums in the table on page 87.

(c) Compare the two sets of figures.

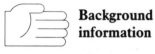

Background information

Classifying dry foods.

Young unripe seeds of garden peas and green beans are still watery. We classify them as Fruits and Vegetables. Frozen, canned and cooked garden peas are still F. & V. but dried peas are not. Peas taste different when they are dried and their food value is different. Dried peas are Fillers. All dried fruits and vegetables are classified as Fillers because they change in flavour and food value when they dry.

Practical activity

Making a pudding with dried fruits.

● Boil some dried fruit(s) in water or orange juice until they are tender.

● Watch them plump up ● You can add spice if you want to but you will not need extra sugar if you use orange juice.

● Record what you did and try to explain why the fruit plumped up and why no extra sugar was needed.

● Think of ways of making your pudding into a square meal.

Discussion: why we buy dry foods.
Consider these points:

1 The kinds of FILLER foods which are dry. Make a list of those which were not originally fruits or vegetables. What were they?

2 How and when do we eat dry foods?

3 How do we store them at home?

4 How long do they keep?

5 How much do they cost compared with fresh foods?

6 How would you store left-over cooked dishes made from dry Filler foods such as rice?

Background information

The storage of dry foods.

Food can become infected with moulds and minute bacteria which can make food go bad. Moulds and bacteria cannot grow without water, so dry foods will keep well. But when dry foods are cooked they absorb water and will obviously then go bad quickly, unless they are frozen.

11

 **Find out** ...   ... about buying dry foods, starting with **PULSES**.

Pulses are ripe, dry pea or bean seeds. There are many kinds of pulses because there are many varieties of peas and beans and also many ways of processing them. Lentils and haricot beans are pulses. Canned baked beans and PROCESSED peas are pulses which are sold ready-cooked and flavoured.

Look in shops and in recipe books and then record your findings about:

(a) The different kinds of pulses we can buy.
(b) The different kinds of pulses stocked by shops in your area.
(c) The names of dishes using pulses – for example Chilli Con Carne.

 Practical activity   Using pulses.

1 Look at a selection of peas, whole lentils and beans that have been soaked overnight in water. Take them apart. Examine them under a low magnification microscope lens. What is the same about all of them – the reason they are all called PULSES?

  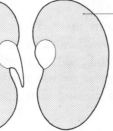 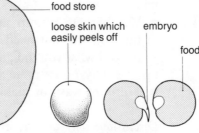 

food store

loose skin which easily peels off   embryo

food store

Butter bean   **Processed pea**   **Brown lentil**

Fig. 9   Pulses examined under a low-powered lens.

2 Discuss with your teacher how to cook pulses.
3 Use pre-cooked pulses to make pastes or pâtés by liquidising or mashing them and adding extra liquid if you want to. Flavour them as you would like. Use canned pre-cooked pulses if you want to as in these examples:

(a) baked beans and spring onions;
(b) kidney beans and yoghurt with chives or chillies;
(c) soya beans and tomato with garlic.

 Extra work   Classifying T.V.P.

Textured vegetable protein is called T.V.P. for short. It is a food made by extracting one part of certain seeds such as soya beans. This part is then flavoured by manufacturers to resemble meat or fish or even blackcurrants. Adding artificial colouring or flavouring to part of a bean does not turn it into meat. So T.V.P. is not a Mains food or F. & V. How can it be classified?

4 Cereals

Cereal grains are the seeds of grasses. They dry naturally on the plant, and as you have already seen, this means they keep well. Some are processed to give us a variety of different foods.

 **Find out . . .**

. . . how many of these kinds of grains you can find in your Home Economics room.

(a) Whole grains such as barley or rice.
(b) Grains broken into pieces – for example, flaked rice, oatmeal or corn meal.
(c) Grains squashed flat and partly cooked – for example, rolled oats.
(d) Grains which have been ground into a powder or flour – for example, whole wheat flour.
(e) Grains which have been ground into a powder but with parts of the grain removed first – for example, white wheat flour or cornflour.

Whatever cereal grain we eat, it is usually mixed with other foods and cooked. How many of the cooked foods listed in this table have you eaten?

GRAINS	PRODUCT	COOKED FOOD
wheat	flour	bread
oats	oatmeal	porridge
rye	flour	crispbread
corn	cornmeal	tortilla
rice	long grain	pilau
barley	pearl barley	soup

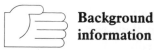 **Background information**

Sort out the different meanings of the word 'CEREALS'.

'CEREALS' is the general name given to the grain crop grown by farmers. 'CEREALS' is also the name given to a type of breakfast food made from grains. Fillers is the name of the food group to which cereals belong. Cereal grains are ripe, dry seeds, with *one* food store (not two food stores as with pulses). The embryo can grow into a plant.

How do you cook with rice?

You must first choose your rice. Rice is a whole grain, the seed of the rice plant. But farmers grow many kinds of rice just as they

13

grow many kinds of apples, for example: long, short and medium grain rices.

The rice may be left untreated or processed in different ways.

Find out ...

... how many different kinds of rice you can find at home or in the Home Economics room.

- Compare them for shape, size, colour and smoothness or roughness. Are they all hard?
- Record your observations.

Practical activity

Cooking different kinds of rice.

- Divide into groups ● Choose ONE type of rice grain.
- Boil 300ml water in a largish pan and add ½ stock cube.
- Weigh out 75g of your chosen rice grain. Add the rice to the pan. Note the time.
- Stir the rice once after the water comes to the boil. Continue to cook the rice slowly.
- Every three minutes, take out a little rice and taste it. Use a clean spoon each time and take care not to burn your tongue.
- Do this until the rice is cooked to your satisfaction.
- Note the time your rice has taken to cook.
- Drain the water off, but collect it and measure how much there is.
- Weigh the cooked rice.
- Display the rice with a label showing: (a) its cooking time; (b) its weight increase; (c) the volume of water drained off.
- Record the cooking times for all the types of rice tried by your class.

Discussion
1 How has the rice changed?
2 What has happened to the water?
3 How do you decide if the rice is done?
4 Do you agree with other groups that their rice is done?
5 Do some kinds of rice stick more easily than others?

Background information

Cooking rice.

Rice absorbs or takes up water and also colours and flavours in the water. People all over the world make use of this idea of **ABSORPTION** to cook rice in interesting ways. You will find examples if you look up Rice Dishes in the index of a cookery book. ('Dish' is the name given to hot or cold mixtures of foods served together.)

Find out ... ... how many different rice dishes your class already knows.

Are any of these on your list?

Risotto   Rice Pudding
Kedgeree   Spanish Rice Salad

 Practical activity   Use the ideas you have been discussing to invent your own rice dish.

- Choose a **SAVOURY** dish (not one with sugar in it).
- Can you give it an interesting name? ● Use 75g of the rice of your choice; 300ml water.
- Remember to make a flow chart before you begin, and ask your teacher to check it ● Use the clues in Fig. 10 to remind you of the questions you should ask yourself:

Fig. 10

- What will you need to make your dish into a square meal?

You will remember that meat, fish, cheese, eggs and milk are all Mains foods. The following table lists some of the Mains foods you could add to a savoury rice dish or serve with one:

Kabanos   Sausages   Corned beef
Fried bacon   Tuna fish   Hard boiled eggs
Cheese   Salmon   Yogurt
Salami   Shrimps

Extra practical activities
1 By now you will have noticed that people from different parts of this country or from other countries may eat rice in different ways. People in India and Pakistan can buy rice which is naturally red or blue in colour ● Investigate how and when people eat rice.

2 Carry out a class survey and find out the details of how and when people eat rice • Ask these questions and add more of your own:

(a) How many times a week do you eat rice (on average)?
(b) At which meal(s) do you usually eat rice?
(c) How do you eat rice? With a fork? With a spoon? With chopsticks?

• Record your findings for (a) so that they are easy to see. One pupil discovered these were the average for her school:

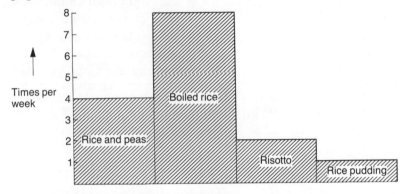

Fig. 11   Number of times per week rice is eaten.

Comparing ways of cooking 'boiled' rice

So far you have cooked rice by adding it to boiling water. Perhaps you cook it in a different way at home. In fact people cook rice in lots of different ways. They each think that their way is best. Find out which you would use and when.

Find out ...

... about methods of cooking rice.

• Work in groups and try out each of these methods and any others your class knows about • Record your results and any problems you have with each method.

1 COOKED DRY – COMPLETE ABSORPTION METHOD
    • Boil ½ cup of rice and 1 cup of water until it has absorbed all the water and the rice is soft.

2 'OVEN' COOKING – ABSORPTION METHOD
    • Cook ½ a cup of rice and 1 cup of water in an oven-proof dish in the oven set at No. 6 or 175°C • Cook the rice until it is done and the water is absorbed.

3 'ON THE MOVE' COOKING – PLENTIFUL WATER METHOD
    • Boil ½ a cup of rice in a lot of water until the rice is done • Drain off the water.

16

- Taste, look at, and discuss rice cooked by each method.
- Which will you use and why? • If you think you will use more than one method, when will you use each? Why?

Rice is only one of the whole grains we use in cookery. Look through recipe books and find recipes using other whole grains. Record the title of the book, the recipe name, and the page number.

Questions
1 Which types of rice would you use to make a rice pudding? Why?
2 Name TWO types of rice you could use to make Pilau rice to serve with a curry.
3 Write a paragraph about the effects of different cooking methods on rice.

What happens to whole grains before they reach the shops? How are they processed?

Harvesting the maize

Cooking with steam in rotating drums

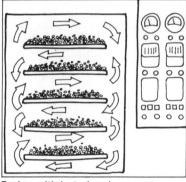
Drying with hot, dry air currents

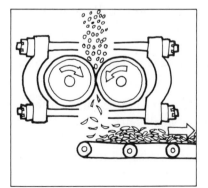
Flaking (squashing into flakes)

Fig. 12

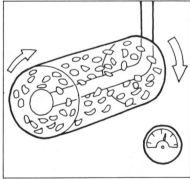
Toasting the flakes to give crispness and flavour

Packaging in air-proof containers

To understand what happens to whole grains after they have been harvested by the farmer, you must see how grains are made up.

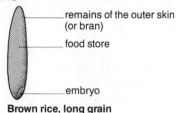

Fig. 13

remains of the outer skin (or bran)

food store

embryo

Brown rice, long grain

• Look at these pictures of rice, and also look at real examples:

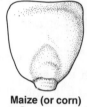

food store

gap left after embryo is removed

White 'polished' rice, long grain

food store

gap left after embryo is removed

White 'pudding' rice, short grain

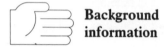

food store

bran

embryo

Wheat

Fig. 14

• Now look at these pictures of other whole grains:

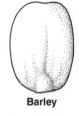

Maize (or corn)   **Barley**   **Oat**

Questions

1 What things are the same for all the grains?
2 What things are different?
3 Where is the embryo in each?

Try to sum up what you have just learned about whole cereal grains. Use this background information to help you.

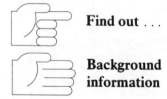

Background information

Cereal grains have a single food store, an embryo plant and a brown coat that is tightly attached and not easily removed. In wheat, when the brown coat is separated, it is called wheat bran. After wheat and rice have been harvested by the farmers, the coat is often removed but the embryo usually comes off too. We call this **REFINING**.

How many other things can be done to grains?

Processed grains have been treated to change them in some way – their shape, or colour, or flavour, for example. Refining is one kind of processing, there are others.

Find out ...

... how many processed grains you have in your Home Economics room?

Background information

They can be squashed like rolled oats or chopped into different sizes from very large pieces to very small ones, or powdered as fine as flour.

The same process is used for corn (maize) to make cornmeal, which is eaten by people in many countries and has different names such as polenta or gramflour.

Grains may also be toasted at high temperatures. This browns them, changes the flavour by 'Maillarding', and makes them very crisp with tiny air bubbles. Some breakfast cereals are toasted in this way.

 **Practical activity**

Make your own MUESLI mix.

● Use pre-cooked rolled grains to make a Muesli mix. You may mix in some oil seeds. Here are some examples:

Rolled oats	Sesame seeds
Cracked wheat	Sunflower
Rolled barley	Melon seeds

You may like to add some of these for extra flavour and colour:

Toasted nuts, oats or wheat, fresh or dried fruits or, a LITTLE sugar.

● Stick examples of all the cereal grains you use on to a page in your notebook and label them ● Record how you made your Muesli and how you could make it into a square meal.

How do we get flour?

From a very early age people have pounded up cereal grains to make a powder we call flour. Any grain can be made into a flour. At first the grains were pounded or ground using two stones or, later, a pestle and mortar. If you look at the picture in Fig. 15 you will see that it must have taken a long time and been hard work.

Fig. 15 Grain being ground between stones on a saddle quern. A rotary handmill is also shown.

 **Practical activity**

● Make your own flour by grinding whole grains of wheat.
● Work in a group and make flour by each of these methods:

1 Using an electric grinder.
2 Using a pestle and mortar.
3 Using the end of a rolling pin on a chopping board. (Put the cereal grains in a strong polythene bag.)

● Put a little sample of each in your folder with an explanation of what you did ● Record which method of making flour made the finest powder ● Sieve the flour carefully ● Look closely at what is left in the sieve. Stick some in your folder. This is mostly

19

the seed coat and the embryo ● Label your sample ● Look at what went through the sieve ● Feel and taste it. This is mainly the ground food store ● Stick some in your folder and label it.

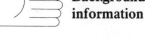

Background information

Different types of flour.

We can get flour from ground whole grains or just the food store. Whole wheat flour and ground rice are two examples. Whole maize grains can be ground up to make gram flour. The ground up food store of maize is called cornflour. It is made up of minute granules called STARCH GRANULES. Custard powder is flavoured, coloured cornflour.

Find out ...

... what cornflour looks like under a microscope.

Under the high power of the microscope it looks like this:

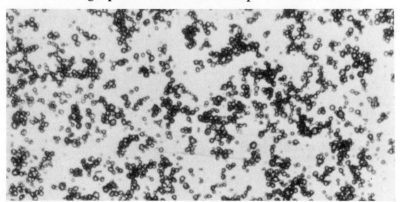

Fig. 16　Raw starch granules.

How do you cook with cornflour?

Find out ...

... how cornflour behaves with water.

1 Put one spoonful of cornflour in a glass with one spoonful of water (use the same spoon for each) and mix well. Label this sample 'A'.
2 Put one spoonful of cornflour in a glass with half a glass of water. Mix well. Label this sample 'B'
3 Stir both very well. Describe what you see. Which mixture is smooth? Which mixture has clumps of unmixed cornflour?
4 Watch what happens and try to describe and explain it.
5 Leave the mixture to stand for 5 minutes. Are the granules of starch heavier or lighter than water?
6 Add enough water to 'A' to make a runny mixture, stirring to avoid lumps. Pour this mixture into a pan and heat it, stirring it well. Watch what happens. Try to describe and explain it.
7 Heat 'B' in a pan, stirring it well. Describe and explain what happens.

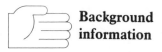
How cornflour behaves with water.

Refined flours have lost the embryo and coat. They are made mainly from the food store. Cornflour is even further refined than white wheat flour. It contains only starch granules.

To make a smooth mixture, make sure that you start with a thick (VISCOUS) mixture. Then when you try to squash out any lumps of cornflour they will not swish away in the watery liquid. You can add extra liquid when all the lumps have gone. Most cooks find it is easier to judge how much liquid to add at first if they ALWAYS ADD THE LIQUID TO THE POWDER. This is because when you dribble or drop the liquid in you can control how much you are adding. The main point is that you must start with a viscous mixture.

How do we make a thick sauce with cornflour?

You are going to make a thick custard sauce for a trifle. If you do not know what a trifle is, look up recipes in a cookery book.

Find out ...

... how much milk you need to make a THICK sauce using 1 tablespoon of cornflour or custard powder. You will find instructions on the packets.

• Find out also how much sugar you need to add to get the sweetness you like.

Discussion
Which flavourings can you add to cornflour if you do not want to make a custard powder sauce for your trifle? You may like to try coffee essence or chocolate powder or lemon essence. Essences are very strong or concentrated, so you will need only a few drops.

Practical activity

Making a cornflour sauce for a trifle.

• Discuss with your teacher the kind of dish to use and the kind of fruit mixture you are going to put at the bottom of your trifle.

• Make a thick sauce using cornflour or custard powder and following the instructions on the packet. As a general guide most people use 1 tablespoon of cornflour and 250ml milk. REMEMBER TO MAKE SURE YOU GET NO LUMPS. In order to make the sauce:

  • Mix the cornflour smoothly with the milk • Add any flavourings • Put everything into a pan and boil the mixture • YOU MUST STIR ALL THE TIME UNTIL THE MIXTURE IS SMOOTH AND THICK • Stir by pressing the mixture against the pan, pressing out any little lumps as soon as they form.

• Leave the sauce to cool a little and then pour it over the fruit mixture. Decorate the top of your trifle with some of the same fruit • Record the changes which you saw as you made the sauce • Use this background information to explain the changes.

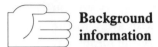

Background information

Starch granules are tiny, hard pieces that sink in water or milk. When cooked they absorb a lot of water, swell up and soften. These enlarged very soft, granules float freely and make the liquid viscous. The process of absorbing water and making a soft jelly is called **GELATINISATION**. Wheat starch will thicken liquids just as corn starch does. Under a high-powered microscope you would see:

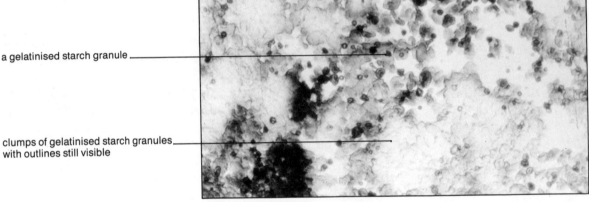

a gelatinised starch granule

clumps of gelatinised starch granules with outlines still visible

Fig. 17 Gelatinised starch granules.

Extra work

You can try putting a drop of water with raw potato starch granules on a slide and then adding another drop of boiled potato starch granules. With low-power there is no need to cover the glass. Use the information on microscopes in the Reference Section to help you.

5 Sugar

Where does it come from?

By now you have used sugar many times in your Home Economics course. You know something about using sugar but how much do you know about sugar itself?

Discussion
What do you know already about sugar and where it comes from? Perhaps someone has lived in a country where sugar cane is grown. Perhaps you know of a farm in this country where sugar beet is grown.

Find out . . .

. . . from books how sugar is made.

- Make a flow chart of the main stages in the process.
- Did you include a stage when sugar crystals are formed? Have you made crystals in Science?
- Did you finish with the stage when the sugar is put into packets?

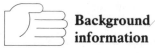

Background information

The cost to health of refined sugar.

We expect to find our sugar in packets and we want it available all the year round. Unfortunately when we gain convenience with foods we often have to change or refine them. We gain convenience, but at a cost. You will have learned a lot about teeth in your primary school so you will already know that there is a connection between sugar and tooth decay. You will study this connection in more detail in Level 3.

Sugar – can you detect it?

If you put a spoonful of sugar on your food you know that you are eating sugar. But what if the sugar is in the sap inside the plant cells, or has been added to cooked foods? Can you tell when there is 'hidden' sugar in food? How good are you at detecting sugar?

Find out . . .

. . . about the taste of sugar.

- Work in two's. The partner who is going to taste should turn around and not look until told to ● Put a small drink of sweetened juice into a glass ● Put the same quantity of un-sweetened juice into another similar glass. The juice must look the same in both glasses ● The partner may then taste and try to spot which is which.

Discussion: the sweetness of juices.
(a) Try to explain why you can not see sugar when it has been added to a liquid.
(b) Which is the sweeter? The juice of ripe or unripe fruit?

Practical activity

Detecting sweetness.

You will need: very sweet biscuits; plain biscuits (unsweetened); a low-sugar scone.
- Let yourself be blindfolded. Your partner will give you two biscuits.
- Taste the two biscuits. One will be very sweet, the other unsweetened. Identify the very sweet biscuit.
- Taste an unsweetened biscuit followed by a low-sugar scone.
- Then immediately taste a very sweet biscuit, followed up by

23

half of the scone. Was there any difference in the way the scone tasted.

• Does it make a difference what you eat just before you taste food?

Find out ...

... which foods contain most sugar.

• Use the food table on p. 86 as you did for watery foods • Record the amount of sugar in 100g of five of your favourite foods • Record the names of the five foods with the lowest amount of sugar • Find five low sugar foods which you like • How much of these do you eat at any one go? We call this amount a **PORTION**.

Background information

Portion sizes.

The size of a portion may depend on how hungry you are, or how much you like the food or whether it comes in a package such as a tube of sweets or a bar of chocolate or a can of baked beans. It could also depend on how the food is eaten. A sandwich has two slices of bread. In this book we have used the idea of an *average* portion to give some guidance as to how much a person may eat at a meal or snack. The sugar content table on p. 86 shows the amount contained in 100g of a food. But we seldom eat exactly 100g. We eat portions. The food table on p. 86 for sugar has some portion sizes for foods, and the number of teaspoons of sugar 'hidden' in that portion. The table has been simplified so all figures are given as whole numbers without any decimal places. One way to find out the sugar content of portions is to draw a graph using the figure for 100g of a particular food, rather as you used graphs to work out costs.

Practical activity

Make a display of food portions showing how much hidden sugar they contain. This is what one class did:

Fig. 18

(a) "7 teaspoons of sugar in this can of Coke".

(b) "25 teaspoons of sugar in this packet of Dolly Mixtures".

When you make your display you can measure out the sugar content in teaspoons rather than weighing it each time.
● Weigh out 25g sugar and find out how many rounded teaspoons of sugar that gives you. We have found that 1 rounded teaspoon of sugar weighs 5g. Do you agree?

Discussion
How does processing affect the amount of sugar we eat.
1 This picture will give you clues:

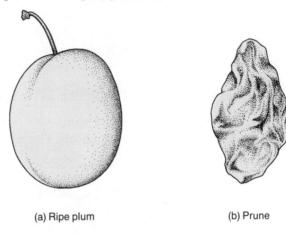

(a) Ripe plum   (b) Prune
Fig. 19

Why does drying or taking away water make the plums sweeter? Have you ever made solutions stronger in Science? Remember the word CONCENTRATED?
2 Make a collection of labels from cans, packets and jars of food. Include soups, sauces, pickles and desserts. Study the list of ingredients on each label and note how many include sugar. Try to think why sugar has been added to foods which we usually call savoury.

Background information

Review of refined Fillers.

1 Bran is the papery casing of cereals. Bran is one form of plant fibre. Many plant foods contain plant fibre, as you will see in Level 3, Food Studies.
2 Sugar is extracted from sugar beet and sugar cane, leaving the rest of the food behind.
3 Oil and fat come from both plants and animals. To obtain the oils and fats the rest of the food is removed.
4 Foods with extra fat and sugar are made-up foods like cakes, sweets and puddings, and fried foods. Bread, some biscuits, and scones are often spread with fat such as butter, margarine or clotted cream.

You have learned something about sugar, so your next Filler food must be FAT.

6 Fat

What does the word mean to you?

If you look back to Level 1, the section on Choosing Foods for Meals, you will remember that the Filler group did not contain only plant foods. Butter is of animal origin but it is a Filler. It is quite easy to eat some fat at every meal but we need to take care not to eat too much. If we eat too much of it, fat is not good for our health. It can make us fat more easily than any other food and it can lead to coronary heart disease. So you really need to know more about fat before you begin to use it in your cookery.

 **Find out** . . .

. . . what the word FAT means to your class.

Fat is a word we use in different ways • How many ways can you think of? The pictures in Fig. 20 will give you some clues:

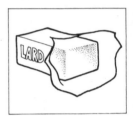

Fig. 20

• Have you any other ideas? • What has a chubby person in common with the things shown in the other pictures? • Check your ideas with this background information.

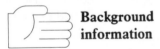 **Background information**

All of us have fatty tissue around our internal organs, such as our kidneys and heart. The kidney shown in Fig. 21 came from a lamb. Your teacher may have bought a lamb's kidney so that you can see how the fatty tissue is wrapped around the kidney. We also have a layer of fatty tissue around our kidneys, but because it is inside our warm bodies it is not solid and hard.

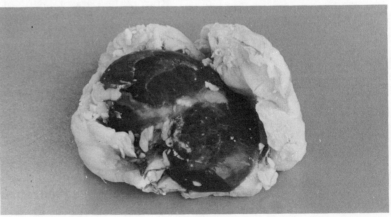

Fig. 21 A lamb's kidney surrounded by fatty tissue.

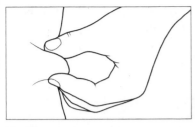

Fig. 22

If you look at a slice of ham you can see that there is a layer of fatty tissue between the rind and the lean part of the ham. You can see a layer of fatty tissue around other meats such as steak.

We also have a layer of fatty tissue under the skin. A fat person has more than a thin person. You can feel it if you pinch your arms and your waist. Fat people have thicker bulges than thin people.

Find out . . .   . . . about body fat.

● Look at these two pork chops.

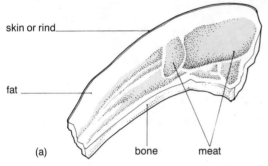

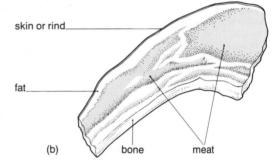

Fig. 23   (a) Pork chop with a *lot* of fat; (b) with a *little* fat.

● Which chop comes from a thin pig and which from a fat pig? Why?

In our bodies some parts are fatter than others. Animals are the same. Look at this picture of bacon. Bacon comes from under a pig's ribs. Along the pig's back, it has a lot of muscle and the bacon is quite lean. Underneath there is more fat and the bacon has lean parts streaked with fat.

Fig. 24

● Look in supermarkets and butchers' shops and find out the names of fatty and lean cuts of pork ● Look up in books or on posters to find out which part of the animal each cut comes from ● Make a note of the cost of the different cuts of pork, lamb and beef.

27

Practical activity

• Look at some fatty tissue from a piece of cooked ham or beef under the microscope • Then look at the fat cells from a piece of raw beef • Record what you see in your notebook. The photograph in Fig. 25 will help you to observe accurately and in detail.

clumps of cells containing soft fat————

thin layer of protoplasm————

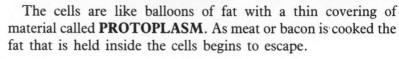

Fig. 25   Microscopic view of fat cells from raw beef.

The cells are like balloons of fat with a thin covering of material called **PROTOPLASM**. As meat or bacon is cooked the fat that is held inside the cells begins to escape.

Practical activity

Investigate the escape of fat for yourself.

• Weigh two similar pieces of bacon and keep a record of the weights.
• Keep one piece for reference and comparison.
• Grill the other piece, using some foil you have weighed underneath it, until it is really crisp.
• Weigh this cooked, crisp piece and record the weight.
• Weigh the foil when it is cold and record the weight.

• Now answer these questions in your book and then discuss them:

(a) Was there a loss of weight after cooking? How much?
(b) What was the weight of fat collected on the foil?
(c) Was it the same as the loss in weight of the piece of bacon? (Think back to what happened to toast when you grilled it.)
(d) Compare the raw and cooked lean bacon. What is different?
(e) Compare the raw with the cooked fatty tissue. What is different?
(f) What happened to the fat on the foil when it cooled?
(g) What is the cold fat called?

Extra practical activity
• Crumble some very crisp bacon and taste it • Describe the flavour. Remember the difference in flavour between fresh and dried fruits? • Try sprinkling crumbled crisp bacon on top of a thick lentil soup or salad. What else would you need for a square meal?

How do we buy fat for cooking?

When you grilled the bacon you caught the hot liquid fat which ran out on to the metal foil. The fat which became a solid when it was cold became a liquid when it was hot. Then when the hot liquid cooled it became solid again. The liquid and the solid were really the same. They were both fat. Some fats are sold as liquids in a bottle. We call liquid fats **OILS**. Oil is merely a fat that is liquid at ordinary room temperatures. If we freeze oil it hardens. If we heat hard fat it softens and becomes oily. We can buy both solid fats and oils.

 **Find out ...**

... about the range of fats and oils in your area.

See how many of these you can find and add others that are not listed.

    Cooking fats – Spry, Pura, Cookeen, White Flora, suet, ghee, white lard.

    Oils – corn, olive, sunflower, safflower, soya, sesame, mustard, ground-nut.

Discussion

Where do you think these fats and oils come from? What is the source of each and how is the oil extracted? Check your ideas with the following background information.

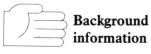 **Background information**

Fats and oils.

Lard is the fat which comes out of the fatty tissue of pigs. Ghee is the pure fat taken from butter.

    Oils are **EXTRACTED** from many plant seeds. Some of these are called oil-seeds for example, sunflower and safflower.

    Corn oil is extracted from the ripened seeds of the maize plant, and soya oil from the soya bean. Oil is extracted from many nuts.

    Cod liver oil is a marine oil (we do not usually cook with it). Oils are extracted from oily fish. Small amounts of fish oils are sometimes used to make cooking fats. These cooking fats are made from a mixture of different oils. They are usually white and have been treated chemically to be hard at room temperatures. Some of the oils have been changed during this process.

    After they have been extracted, oils are purified. This process is called refining. Look back to the work you did on sugar. Extracted and refined oils, like sugar are Filler foods, even when the fat or oil has come from animals. The lean meat or fish left behind is still a Mains. In the case of soya beans and ripe corn seeds the foods left behind, after the oil has been extracted, are still Fillers.

29

. . . what we buy when we buy butter and margarine.

- Melt 100g butter and 100g margarine, but do not over-heat.
- Pour each carefully into a 100ml cylinder and leave to cool.
- Look carefully and record what you see. You should see something like this.

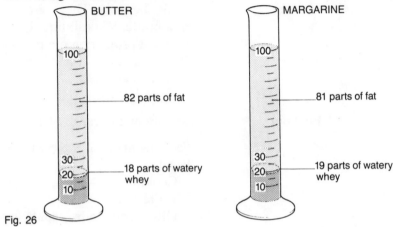

Fig. 26

- Now do the same for a low-fat spread and a soft margarine.
- Record all your results, and then make bar graphs showing how much fat was present in each sample.

Discussion

1 How much water was there in each sample?
2 What was the white mixture which sank to the bottom of the cylinder?
3 Why should anyone want the pure fat above the white mixture in the cylinder? What is the layer on the top?

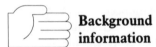
Background information

Ghee.

Ghee is made from butter. Clarified butter is the same thing as ghee. This is made as you have just done. The butter is heated gently to liquify it and then it is left to settle and cool a little. Salt and other solid ingredients which give butter some of its flavour settle to the bottom. Then the layer of fat can be taken off. This layer is pure fat.

In hot countries, before refrigerators were common, people found that ghee kept better than butter. It was also better for frying. The water in butter splutters when it is heated and the solids tend to make food stick to the pan. Ghee is pure fat: it contains nothing else.

Extra work

Try to find advertisements where something is described as PURE. Discuss how and why the word PURE is used in each advertisement.

How does butter compare with milk?

If you look in books you will find the details of how butter is made from milk. These bar graphs tell you about the amounts of fat and water each contain.

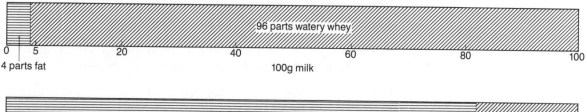

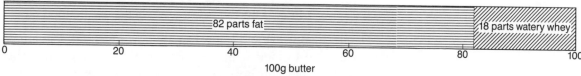

Fig. 27   Fat and water-content of milk and butter.

Questions

1 How much butter would you have to use to make 100g of ghee?
2 Are butter and milk the same sort of food?

• Look back to the food groups in Level 1. You will find that milk is in the Mains group, but butter is in the Filler group. Can you now work out why they are in different groups?

What are the differences between one kind of margarine and another?

One kind of margarine is hard and comes in a block wrapped in foil or waxed paper.

Soft margarines are of two kinds. They both come in tubs and look the same. They may taste very similar but they are made from different types of oils. To identify them look on the top of the tub. If it says: 'HIGH IN **POLYUNSATURATES**' or HIGH IN '**POLYUNSATURATED FATTY ACIDS**' it is one type. If the tub does not have this label, the margarine is ordinary SOFT MARGARINE. Low fat spreads look like margarine and they are made to taste very similar. But they are at least half water – that is why they are 'low-fat'. They can not be used for cooking.

Checking for hidden fats

There are lots of made-up meats which have ground-up fatty tissue added to them. You can often see the fat in sausages and the lumps in black puddings. Sometimes fat is added to lean, ground meat in such a way that it can not be seen. Luncheon meat and some pastes are sold like this. These foods are FATTY MAINS (doubles).

Find out . . .   . . . how to discover if foods have fat in them or not.

• Study this list:

beefburgers; luncheon meat; roasted peanuts; wholemeal bread; crisps; sausage rolls; Cottage cheese; Cheddar cheese; mayonnaise; salad cream; single cream.

• Make a list of the foods which you think contain fat and say why you think they do. Some of the reasons one pupil gave were:

'They feel greasy' . . . . 'They are fried' . . . . 'I looked up the recipe' . . . . 'I looked up the list of ingredients on the food label'.

To check your list and know exactly how much fat is in a food you will have to look it up in a food table. None of the other methods can tell you how much fat is present. Even the list of ingredients on the food label on a packaged food does not give you quantities. But remember that if the fat or oil comes near the top of the list it means that the food contains a lot of fat compared with most of the other ingredients.

Practical activity   Working out the amount of fat in portions of food.

• Using the food table on p. 87 work out the amount of fat, to the nearest gram, in the following foods:

100g Beefburger (fried)	80g Sausage rolls
50g Luncheon meat	40g Cottage cheese
50g Peanuts (roasted)	45g Cheddar cheese
50g Wholemeal bread	25g Mayonnaise
25g Potato crisps	25g Salad cream
25g Single cream	

• Were you surprised at any of the foods? Why?
• Using liquid oil make a display of the different amounts of fat in each food. If you use a glass and oil as shown in Fig. 28, you will be able to compare the amounts of fat easily.

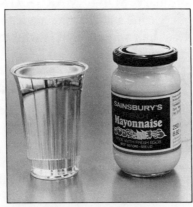

Fig. 28   Mayonnaise: 198g of fat in 250g.   Peanuts: 49g of fat in 100g.

• Record this information in your book for future reference.

Extra work

Decide how much you eat of each of five of your favourite foods at any one meal. Then calculate the amount of fat present in a portion of each of the five foods.

If your favourite foods are not included in the table, discuss the matter with your teacher. You may be able to find a similar food or you may have to use another food table.

Are you in charge of the fat in your diet?

Some fat people eat very little. They are just not able to burn off the extra food they eat. However, most fat people, and there are a lot, would probably lose body fat just by cutting down on the amount of fat they eat.

So, for most of us, the less fatty foods we eat, the better for our figures and for our hearts.

Discussion: What emotions are these people feeling?

Fig. 29

33

7 Flour mixtures

How many different kinds of flour mixtures are there?

Of course it all depends on what you mean by 'flour mixtures'. To begin with we are going to think about mixtures which contain wheat flour. We are also going to think only about mixtures which are dry when we eat them. Anything liquid, like soups and sauces, will have to wait for a while.

 **Find out** ...

... how many baked goods you already know which contain wheat flour.

Everyone will be able to think of BREAD ● Make a list for yourself and then compare it with a partner's ● Now try to describe everything you can remember about ONE food on your list ● Read the descriptions in Fig. 30 carefully to see how to describe your baked goods **SYSTEMATICALLY**.

FAIRY CAKE

Shape: small, domed, round shape.

Top: brown on top and crisp.

Sides: paler brown where the paper cup has been in contact with the cake. Softer than top.

Inside: pale inside and spongy; crumbles when you bite it; has holes inside where there were bubbles.

Fig. 30

You may not agree with all the words used in these descriptions. Discuss them with your teacher and the rest of the class. Did you find any foods on your list which fitted these descriptions? Did you find more than one food which fitted any one of these descriptions? If so, you are starting to classify flour goods. You are beginning to find out how many different kinds there are.

You are going to start your investigation of flour goods by looking at three kinds:

1 spongy/bubbly and crumbly: fairy cake.
2 crisp and crumbly: shortbread.
3 flexible/bendy and spongy: bread.

Practical activity

● Look up recipes for these goods in your cookery book.
● Note the headings they come under in the index of recipes.
● Note the other ingredients listed in each recipe ● Which ingredients are the same for all recipes?

Discussion: what is a cake?
Is gingerbread a cake? Are pastries cakes?

What happens when we mix flour and water and heat them?

Most of the flour goods you are going to make have some liquid in them, and all have to be cooked in some way. So your investigation of flour goods must start by mixing flour and water and heating them.

Find out . . .

. . . the results of mixing flour with hot water.

Do you remember how to mix a starch powder with a liquid so that there are no lumps? ● Use the same idea to mix the flour and water.
● Make two fairly thick mixtures in a basin using:

(a) 1 tbsp. of white plain flour and 2 tbsp. of cold water.
(b) 1 tbsp. of white plain flour and 2 tbsp. of boiling water.

● Record all the changes in each mixture, including changes in flavour ● Note particularly any differences. If you have time, cook each mixture by any method you choose, but use the same method for each ● Repeat this test using self-raising flour (S.R. flour for short). Again, if you have time, cook each mixture, using the same method of cooking as before.

Discussion: flour and water mixtures.
1 What effect did the hot water have on the mixtures?
2 How was the S.R. flour different?
3 Can you think of any ways of preparing and cooking the mixture so as to make it crisp and browner?

4 What could you add to improve the flavour?

5 If you cooked and tasted them, which of the mixtures did you prefer?

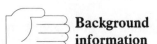

Background information

Cooked flour mixtures.

You probably found that cooked plain flour and water was not very appetising. In fact the uncooked mixture can be used as a glue. However, when people had little else to eat they found ways of making a flour and water mixture into something they could enjoy eating.

The secret is getting some gas bubbles into the mixture so that it is not so solid. One way is to use self-raising flour.

When you mixed plain flour with water you probably mixed in a little air, accidently, but there were very few bubbles. Air is a gas just like steam. You watched the bubbles of steam forming in the boiling water (Level 1) and rising to the surface. But when you used self-raising flour there were many more bubbles. This was because of the baking powder which had been added to make self-raising flour. It gave off bubbles of a gas called **CARBON DIOXIDE**. You have met this gas if you have drunk a fizzy or carbonated drink.

The baking powder in self-raising flour only gives off its bubbles if it is wetted and heated.

Practical activity

Using cooked flour and water mixtures.

You will need some biscuits such as Water biscuits (or you may know of others). Water biscuits are made from flour and water, flavoured with salt, and rolled or shaped very thinly and baked or griddled until they are crisp and brown. In Britain biscuits are usually baked. Water biscuits can have blisters as well as bubbles inside them. They are not easy to make, but you can buy them in packets.

- Taste one and describe what it is like in your notebook.
- Then work with a group and try putting different toppings on the biscuits.
- Think about colour, shape, flavour and how it feels in the mouth.
- If possible leave out some biscuits with moist toppings.

Find out . . .

. . . what happens to a crisp biscuit which has a moist topping left on it.

- Put some topping on one half of a biscuit. Leave the other half plain. Look and feel it every 5 minutes. Note any changes.

Perhaps someone in your class can tell you about other cooked flour and water mixtures such as Chapati.

How can we make a flour mixture which is flexible, soft inside, and tastes good when cooked?

You have seen that self-raising flour produces a gas which makes bubbles in a flour mixture. Now you are going to investigate what happens to those bubbles. To do this you will have to think about other things at the same time. You are going to make and cook a flour mixture containing several ingredients. When you cook the mixture, the heat will cause a lot of changes all at the same time. You are going to be busy! You are going to make SCOTCH PANCAKES.

Find out ...

... what a Scotch pancake is like.

● Look at coloured pictures of Scotch pancakes in a cookery book. Better still, look at a real one if your teacher has been able to provide one for you ● Describe how it looks, feels, and tastes. Compare your description with the ones in Fig. 31. You may want to use different words which make more sense to you. Discuss the words you choose with your class.

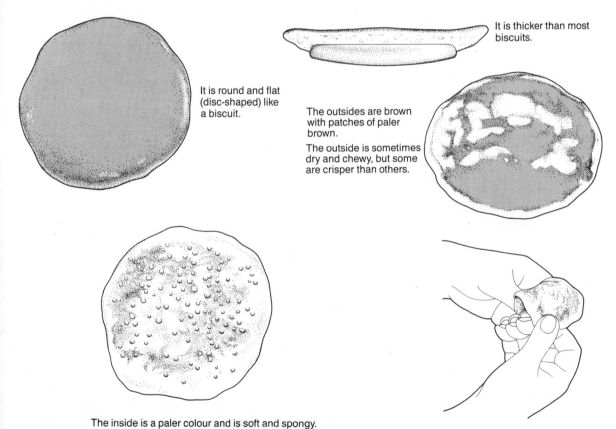

It is thicker than most biscuits.

It is round and flat (disc-shaped) like a biscuit.

The outsides are brown with patches of paler brown.

The outside is sometimes dry and chewy, but some are crisper than others.

The inside is a paler colour and is soft and spongy.
There are long tunnels running up from the bottom.

It is fairly flexible. It bends without breaking.

Fig. 31

37

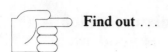

Find out ...   ... how each ingredient helps to make Scotch pancakes.

● Look at the recipe and think back to the work you did in Level 1. Then see if you can work out how to get the six points we used to describe the cooked Scotch pancakes. The background information will help you to decide what each ingredient does.

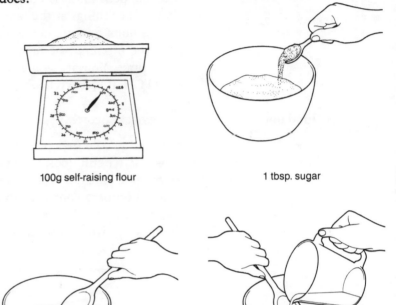

100g self-raising flour   1 tbsp. sugar

Fig. 32   Ingredients for Scotch pancakes.   1 egg   150 ml milk or milk and water.

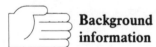

Background information

Scotch pancakes.

FLOUR provides the bulk in many flour mixtures. A fairly runny mixture of flour and water can set firm if it is cooked. The starch granules from the ground-up food store of the wheat grain absorb water and gelatinise during the cooking. The outside can dry out and make a crust at the same time. This crust can brown because it is heated strongly. (Think back to Maillarding in Level 1.)

SELF-RAISING FLOUR makes bubbles of gas when it is mixed with water and heated. These make the mixture spongy but they also take up a lot of room so the dough gets bigger; rather like blowing up a balloon. We say the dough has been **AERATED**. The tunnels in a cooked pancake are left after some of the gases have expanded and escaped during cooking. If the gases are trapped the bubbles remain.

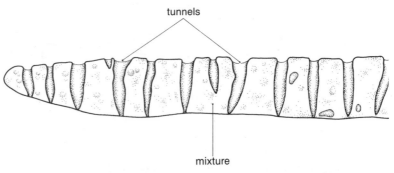

tunnels

mixture

Fig. 33 Cross-section of a Scotch pancake showing the tunnels that the expanding air bubbles made in the mixture.

SUGAR obviously makes the pancakes taste sweeter.

EGG is liquid. It is stickier and thicker than milk or water and helps to BIND the mixture together. Egg coagulates when heated so it helps to set the pancake, including the walls between the tunnels. The egg helps the mixture to brown well on the outside during cooking. The milk helps too and also adds flavour.

Scotch pancakes are sometimes called drop scones. You may find it confusing that one recipe can have two names. What are they called in your area?

They are not at all like other scones. Scones are made from a much thicker dough and they do not usually contain egg.

You will come across many examples of one mixture being called by different names in different areas. It can be fun to try to find out the reasons behind the names.

You now know the recipe and you could try to follow the instructions for making Scotch pancakes in a cookery book. But this is the first mixture you have cooked in this course. You will feel far more confident if you can work out for yourself what you should do. Use the books to check your ideas, after you have made your decisions.

Discussion
Making decisions about the method to use for cooking Scotch pancakes.

● Work with a small group ● Keep thinking about what a cooked Scotch pancake is like ● Try to decide how you will get the results you want. The sort of decisions you will have to make are shown in Fig. 34.

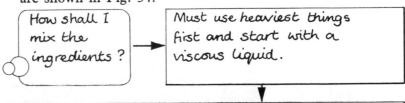

How shall I mix the ingredients?

Must use heaviest things first and start with a viscous liquid.

Decision: I'll mix all the dry ingredients first and then add the liquid slowly.

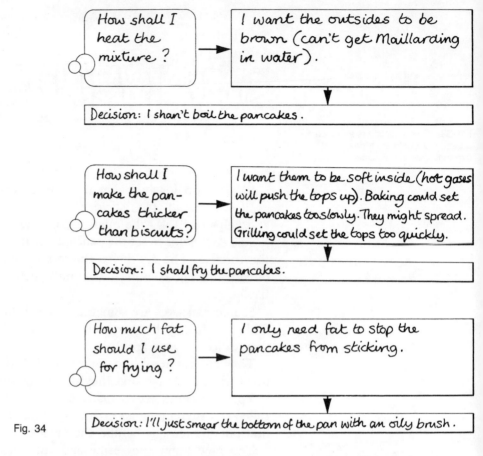

Fig. 34

Other decisions you will want to make will include:

(a) What will I use to get the same amount of mixture into the pan each time?
(b) Do I want the pancakes to be round or oval? How do I get the shape I want?
(c) How can I know how hot the pan should be?
(d) How shall I turn the pancakes over so that the top gets cooked?
(e) How 'done' do I want them? How brown do I want them to be?
(f) How shall I know when they are done?

Practical activity   Making Scotch pancakes.

● Make a flow chart to plan the order of work ● Make some Scotch pancakes and see if yours match the description given earlier in this section ● If some do not, try to explain why.
● Enjoy your drop scones! Are they –

FLEXIBLE?
SOFT INSIDE?
NICE TO EAT?

Are there other mixtures like Scotch pancakes?

You have just made what is called a **BATTER**. There are different kinds of batter. Some are thick and some are thin. You fried your batter to make Scotch pancakes. You dropped spoonfuls of the batter onto the pan and perhaps were able to cook several at the same time. You could have made one big pancake or you could have baked the batter. The results would have been different.

 **Find out ...**

... ways of using an egg and flour batter.

● Look up recipes for batters ● Record the names and the way they are cooked ● Also record the name of the cookery book and the number of the page.
● Note any variations from the Scotch pancake recipe. Can you judge which will be runnier and which will be stiffer?
● Make a table like the following one in your book. Two examples have been filled in to show you what you have to do.

Batter recipes showing variations from the Scotch pancake recipe.

Recipe	Difference in Ingredients	Difference in Quantities	Differences in Shape	Different ways of making
Pancake	no sugar	250ml milk (more liquid)	large flat disc	pour from jug
Yorkshire pudding	no sugar	250ml milk (more liquid)	round or rectangle– well risen	baked

 Practical activity

Making pancakes.

● Follow the recipe and method in a cookery book for making pancakes. If your pancakes do not turn out as you want them to at first, try to think why. They will be successful if you take care at every stage and understand what you are trying to do.
● If you have time make different fillings for your pancakes.

Biscuits, pancakes, bread – what are the differences?

The Water biscuits were crisp, dry, blistered and thin. The Scotch pancakes were softer, smoother and thicker. A loaf of bread is obviously bigger than the biscuits or pancakes. It has a lot more 'inside', (crumbs) and not so much 'outside', (crust). But it is not easy to describe bread because there are so many different kinds.

Find out . . .   . . . how many different kinds of bread your class knows.

- Use the photograph in Fig. 35 to help you to decide what you mean by the word 'bread' ● Is all bread soft? ● Or is some bread crisp? ● How many shapes of bread do you know? ● Does all bread taste the same?

Fig. 35

- Copy this table into your books and then complete it, adding examples.

Examples of different breads	Shape and size	Flavour	The outside	The inside	How it is used and eaten
white, sliced	large rectangle	bland	brown, tough	white, spongy and chewy	with something spread on it

- Which ingredients have to be mixed together to make bread? You should find labels on wrapped bread ● Find out which foods come first on the list of ingredients ● Look up recipes for making bread.

Discussion

What is the job of each ingredient in a bread recipe? You could make bread with these ingredients in these quantities:

500g plain flour (white or wholemeal)
15–20g fresh yeast
3 tsp salt (optional, used to be essential)
1 tsp sugar (optional)
300ml warm water. (30°C)

- Try to work out what job each ingredient does ● Think about:
(a) bulk; (b) aeration; (c) flavour; (d) the feel of bread in the mouth.

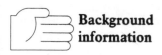
Background
information

YEAST is a microscopic plant. You can see yeast if you mix a little fresh yeast with water and smear it on to a microscope slide. If the single-celled plants are warm, have food and water, they are very active. Then they give off carbon dioxide gas. This is used to aerate the kneaded bread dough during the stages before the dough is cooked.

Find out . . .

. . . about yeast from books and material produced by the Flour Advisory Bureau in London. Companies selling bread or flour and yeast also produce information leaflets.

Background
information

Dough and batter mixtures.

One difference between the recipes for Scotch pancakes and bread is the amount of water. Compare these amounts:

Scotch Pancakes	Bread
100g flour	500g flour
150ml water	300ml water

To make the comparison we must make the weights of flour the same:

Scotch Pancakes	Bread
500g flour	500g flour
750ml water	300ml water

The pancake mixture contains more water than the bread mixture. The **RATIO** of water to flour is bigger for the pancakes. Because the pancake mixture contains a lot of water it is a runnier mixture than the bread mixture. It has a runnier consistency. The pancake mixture is called a batter. The bread mixture is called a **DOUGH**.

A dough mixture is stiffer than a batter because it contains less liquid. It has a stiffer consistency.

Discussion
Why is a loaf of bread baked and not fried?

Think about the differences between Scotch pancakes and bread, for example size, shape, thickness, their outsides, their insides, etc.

What happens when we add fat as well as bubbles to a dough?

The cooked mixtures you have investigated so far have had either no fat or very little fat in them. Yet even with only a few ingredients so many varieties are possible. Think of the many different kinds of bread you know about. Think about the different quantities of ingredients. Remember how the doughs were baked in different ways. You are going to need to remember all these things when you investigate SCONES. You had better find out just what a scone is before you make one.

 **Find out . . .**

. . . what is a scone?

● Look at real scones and also feel and taste them.
● Copy this chart in your book and then complete it for as many scones as possible.

Type	Colour	Shape	Flavours	Use
plain, sweet	pale golden	small round	sweet	with sweet spreads

All your examples are called scones. Yet they look and taste different ● What do they have in common? ● Look up recipes and find out which ingredients are mixed together to make scones.

Investigating ways of cooking scone dough

Your teacher will have made up some scone dough. The ingredients have been mixed already because you are going to concentrate on what to do to the raw dough.

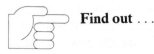 **Find out . . .**

. . . what must be done to raw scone dough to make a cooked scone you will like.

● Look at and feel the raw dough and try to describe it.
● Look at and feel a cooked scone and try to describe it. This will give you an idea of the kind of scone one person liked.
● Now work with a partner and carry out these tests. Make sure all the tests are done by different groups in the class.
● First check with the background information what is meant by a **FAIR TEST**.

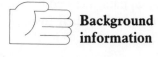 **Background information**

Fair tests.

So far when you have done an investigation you have either observed something or done a test on two different things and then compared the result. In order to investigate ways of cooking scone dough you will need several different tests and you will need to consider results from all of them. Each group will compare two scones that have both been made in exactly the same way except for ONE difference. For example, both are exactly the same except that one set is rolled out THICKER. Tests like this are called FAIR TESTS.

 **Practical activity**

Carry out fair tests to find out what happens when you cook scones in the following ways.

1 *IN DIFFERENT OVEN ZONES*
   • Roll out the dough to 2cm thick and cut out 3 scones, using the same cutter • Bake one scone on the top shelf, one on the middle shelf, and the last one on the bottom shelf • To make the test fair bake all the scones for the same time – 10 minutes.

2 *FOR DIFFERENT COOKING TIMES*
   • Roll out the scone dough to 2cm thick and cut 3 scones, using the same cutter • To make the test fair put all the scones in the same oven, on the same shelf • Remove one scone after 5 minutes • Remove another scone after 10 minutes, and then remove the last scone after 15 minutes baking.

3 *USING DIFFERENT THICKNESSES OF SCONE DOUGH*
   • Using the same cutter, cut out 3 scones • One must be 1 cm thick, one must be 2cm thick and the third 3cm thick • To make the test fair bake all the scones on the same tray, in the same oven for 10 minutes.

4 *BRUSHED WITH DIFFERENT GLAZES*
   • Roll out the dough to 2cm thick, and cut out 3 scones using the same cutter • Glaze one scone with beaten egg, one with milk, and leave the last one plain (no glaze) • To make the test fair bake all the scones on the same tray, on the top shelf, for 10 minutes.

5 *USING DIFFERENT SIZED CUTTERS*
   • Use scone dough rolled out to 2cm thick, but one scone should be cut with a large cutter, one scone with a medium-sized cutter and use a small cutter for the last scone • To make the test fair bake them on a tray, on the top shelf for 10 minutes.

   • Make a check list of points which will tell you when the kind of scones you like are 'done'. One pupil started like this – 'MY SCONES WILL BE DONE WHEN THEY MOVE IF I SHAKE THE TRAY' • Is that right? Why?
   • Look at and taste the results from all the tests • Decide which scones you like best • Record your choices.

Now you must find out how to mix the ingredients for scones together.

How are the ingredients for scones mixed together?

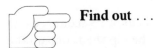 **Find out ...**

... what you need to know before you can decide how to mix the ingredients for scones.

1 Look at the ingredients listed in one scone recipe:

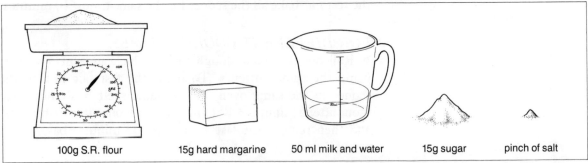

100g S.R. flour   15g hard margarine   50 ml milk and water   15g sugar   pinch of salt

Fig. 36 Ingredients for a scone recipe.

2 Decide which of the ingredients you can mix together because they are all dry. Remember making Scotch pancakes!
3 Investigate the problems of mixing hard fat with flour.
 • Work with a partner and try to think out the answers to these questions:

 (a) How can you spread out the fat evenly into all parts of the flour? Would it help to cut the fat into small pieces? Should you squash it? Gently? Roughly?
 (b) Which tools will help you to mix-in the fat? A fork? A spoon? Your hands? Your fingers?
 (c) How will you tell when it has been mixed in enough?
 (d) Which methods of mixing fat with flour have you already seen, perhaps at home? Do they work well?

4 Decide which method you think will work best and why.
5 Compare your decision with other groups in your class.
6 If possible try out some of your ideas in practice with your teacher.
 • How would the method of mixing be different if you used soft fat?

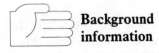 **Background information**

Mixing fat with flour.

Hard fat must be cut into small pieces and gently squashed or RUBBED IN to the flour until no lumps can be seen, even when the mixture is shaken. Some cooks rub the mixture gently between the palms of their hands. Some use their fingers. Hard fat is mixed in before the water is added because fat does not mix easily with flour if there is a lot of liquid around.

   Soft fat can be mixed just by stirring it in with a fork, even if the liquid is put in at the same time.

Practical activity   Making scones as you like them.

1 Decide exactly what kind of scone you want to eat. Make a list of the important details you want to achieve. Make sure you include details about SIZE, COLOUR, FEEL and FLAVOUR.
2 Write out a check-list for testing when your scones are cooked or 'done' as you want them to be. The fair tests will help you.
3 Decide which ingredients you are going to use. You may want to add extra ingredients for flavour. Remember your basic ingredients were:

> **Ingredients for Scones**
> 100g S.R. flour   15g sugar (optional)
> 15g hard fat   pinch of salt
> 50ml milk and water

4 Decide all the details of the methods you are going to use, remembering your fair test decisions. Work out your flow chart and have it checked.
5 Make your scones and wash-up.
6 Eat your scones and decide if you have been successful. Were they the right size, colour, feel and flavour?
7 Record what you did and the results. Suggest, with reasons, any changes you would make next time you make scones.

How much do scones cost?

Find out . . .   . . . how much your scones cost.

• You will need to know the cost of the foods as follows:

  (a) 6 eggs cost ....
  (b) 500ml milk costs ....
  (c) 500g hard fat costs ....
  (d) 500g sugar costs ....
  (e) 500g self-raising flour costs .....

• Then make costing graphs for each so that you can work out the costs of the ingredients you used.
• Add up all the costs and record them like this:

Quantity	Foods	Cost
	Total cost =	

Questions
1 How many scones did you make?
2 How much did one scone cost?
3 How much does ONE shop-bought scone cost? (It must be the same size.)
4 How much do eight packet-mix scones cost?

Pastry: what is it and what is it used for?

 Find out . . .

. . . what pastry is used for.

● Make a list of all the ways you can think of that pastry is used. For example:

(a) as a container for wrapping other foods in;
(b) to eke out expensive foods and make the dish cheaper per portion.

You can see from this that pastry has many different uses. Each kind of pastry has its own special uses. For the moment we are going to concentrate on **SHORT CRUST PASTRY**. You are going to find out just what it is, and what it is used for. You will start, as you did with scones by getting used to the feel of the uncooked dough.

 Find out . . .

. . . about the consistency of short crust pastry dough and how to roll it.

One of the difficult things about making dishes using short crust pastry is getting the pastry into the shape you want. There are tricks to this, but the most important thing is the consistency of the raw dough. You have to learn what it feels like.

● Use some uncooked prepared pastry. You can use a packet of frozen bought pastry or your teacher may have been able to make some up in advance, using a food mixer and perhaps storing it in a freezer. The dough must be at room temperature.
● Pull off a little piece (about 30g) and get used to the feel of it before you start rolling out the main piece of dough.
● Divide the small piece of dough into two pieces.
● Add some flour to one piece.
● Feel each piece; pull each piece; try to squash out each piece on the table with your flat hand.

● Now answer these questions:

(a) Which word would you use to describe each piece? Sticky? Dry? Soft? Firm? Stretchy?
(b) What did adding flour do?

(c) Which piece will be easier to roll? Why?

(d) Did either piece extend as much as bread dough before it snapped?

 Practical activity Making a quick quiche.

You are going to use the main piece of dough to make some quiches. Fig. 37 shows one suggestion for a quick quiche:

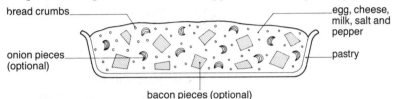

bread crumbs

egg, cheese, milk, salt and pepper

onion pieces (optional)

pastry

bacon pieces (optional)

Fig. 37

The pictures in Fig. 38 show you points to think about when making your quiches. Note that you have to roll out your dough thinner than you did when you made scones. Keep all your trimmings after you have cut out your circles. You will need them for your next investigation. Work out a flow chart and then make some quiches.

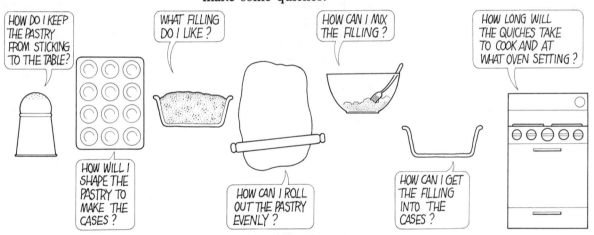

HOW DO I KEEP THE PASTRY FROM STICKING TO THE TABLE?

HOW WILL I SHAPE THE PASTRY TO MAKE THE CASES?

WHAT FILLING DO I LIKE?

HOW CAN I ROLL OUT THE PASTRY EVENLY?

HOW CAN I MIX THE FILLING?

HOW CAN I GET THE FILLING INTO THE CASES?

HOW LONG WILL THE QUICHES TAKE TO COOK AND AT WHAT OVEN SETTING?

Fig. 38

Make sure your flow chart includes a check list for 'doneness'.

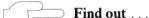 **Find out** . . .

. . . more about rolling pastry.

● Use scraps of pastry left over from making the quiches.

● Work in small groups and invent some tests to investigate the following:

(a) The different ways of rolling out a ball of pastry evenly so that it ends up as a sheet of the thickness you want.

(b) The different ways of making a circle without a cutter.

(c) Ways of making a square without any wastage.

(d) Using a lot of flour and a little on the table when rolling out.

(e) How much water you need to join two pieces of pastry together.

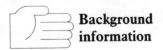

Background information

Rolling pastry.

The pastry you used had just the right consistency. It was not so sticky that it was difficult to lift off the table on the rolling pin. And it was not so dry that it cracked when it was rolled out.

You probably found that if you put too much flour on the table your pastry moved back after you rolled it out. It is better if it JUST sticks to the table. Pastry can only be stretched out a little way at any one time. But after a short rest it can be stretched a bit further.

Find out . . .

. . . about the ingredients used for making short crust pastry.

- Compare the ingredients for making pastry and scones.
- What is the same? What is different?
- Which is wetter? Which is fattier?

Background information

The ingredients used for short crust pastry.

FLOUR is the chief ingredient. When it is mixed with a little water and heated it sets and holds its shape.

WATER binds or sticks the mixture together. Too much water makes the cooked pastry hard though it makes the dough softer. You can try a comparative test for this. But make it fair.

FAT is spread evenly through the mixture and tenderises the pastry. When the fat is being mixed in, air sticks to it. The air bubbles stick to the fat as it spreads through the flour mixture, so fat adds air to pastry. These air bubbles expand helped by steam when the pastry is cooked making the pastry crumbly, so it breaks more easily in our mouths. Otherwise it would be hard. Some cooks put extra bubbles in pastry by using self-raising flour. If we use self-raising flour we do not need so much fat.

If you have time you can investigate these variations by making some comparative tests.

Find out . . .

. . . if you can make short crust pastry.

- Work out strategies for both hard and soft fats.
- Decide which kind of fat you will use and a recipe suitable for that fat. (Look up recipes in cookery books.)
- Now see if you can mix the ingredients yourself and make a dough just right for rolling.
- Work out all the questions you will need to ask yourself. Here are some clues:

(a) Have I measured the ingredients accurately? (b) Are there any lumps of fat showing? (c) Do I need one more drop of water?

- Compare your pastry with a partners. Do you think yours will roll well? • For now, cover your pastry to prevent it from drying • Label it with your name • If you are not going to use it until the next lesson you will have to freeze it. Your teacher will organise the freezing.

 **Practical activity**   Making a pastry dish.

1 Planning

When you were making scones you decided on the size, brown-ess, etc., of the scones you wanted to make. We say you decided on the **CRITERIA** for your finished scones. This time we will give you the criteria for a pastry dish:

(a) PURPOSE: the pastry must act as a container;
(b) COST: the pastry must make the filling go further (you will want to eat less so it will cost less);
(c) TIME: the dish will be finished in the time set by your teacher;
(d) HEALTH: the dish will be low in fat and sugar but do not try to reduce the fat in the pastry;
(e) SKILLS: the dish will involve some preparation of fruit and/or vegetables;
(f) SUITABILITY: the dish can be carried home safely.

You are going to work in groups organised by your teacher.
• Decide on the dish you are going to make • Work out the ingredients you will need • Arrange to get the ingredients.
• Show that the dish will meet the criteria • When your criteria have been checked by your teacher work out the stages of the method, that is all the jobs you must do • Make a flow chart so that you know the order in which to do the jobs • Include a reminder to turn the oven on and set it for pastry.

2 Carrying out your plans.
• Make your dish • Enjoy your finished dish and decide if your pastry meets your criteria • If not, decide how you could improve things.

What have you learned about flour mixtures?

Questions

1 Look at these recipes then answer the questions:

A	B
200g S.R. flour	200g S.R. flour
2 eggs	30g margarine
50g sugar	30g sugar
300g milk/water	100ml milk/water

C	D
200g flour	200g Plain flour
100g marg./cooking fat	2 eggs
20ml water approx.	500ml milk/water

(a) Which is a scone mixture? Why?

(b) Which is a pancake mixture? Why?

(c) Could you use recipe A as a pastry? Why?

(d) Which mixtures could you roll? Why?

(e) Which mixtures would rise the most when cooked? Why?

2 You are bored with using pastry for pies. What could you use instead?

3 You want to make cheese scones and fruit scones at the same time. What is the quickest way to do this?

4 Some ingredients are vital in a recipe. You cannot make scones without flour. Other ingredients can be varied. Invent four new recipes for scones by choosing different 'variable' ingredients.

5 You have made 8 scones with this recipe: 100g S.R. flour; 15g margarine; 15g sugar; 50ml milk/water. What would you need to make 16 scones?

What have you learned about classifying foods, analysing meals and portions and checking foods for fat and sugar?

 **Extra work**

1 Classify the foods included in these meals in the 4 groups – remember about doubles and triples.

A	B	C	D
sausages	orange juice	buttered scones	fried cod
chips	boiled egg	cream cakes	chips
jelly	wholemeal bread	tea	peas
	milk		rice pudding
			coke

2 Decide if each meal is a square meal. If not, suggest some foods which could be added to make it square, or foods which could be exchanged for one already there.

3 Invent two meals which are not square. Swop yours with a partner's. Square your partner's meals so that you would like to eat them.

4 Using foods that are included in the food tables, plan meals which have:

(a) high fat or high sugar or both;

(b) low fat or low sugar or both.

See if you could estimate the amount of fat or sugar if the food has not been included in the food tables.

Check for a lot of Fruit and Vegetables and unrefined Fillers.

5 Collect wrappers from manufactured foods. Check them for quantities and ingredients. How many portions do each give?

6 Draw some bar graphs for the fat and water content for foods from each of the four food groups. Remember that watery foods and all low-fat foods help slimness.

7 Use the food tables to help you to decide on a list of alternatives to high fat biscuits and high sugar sweets. Then try to find some alternatives to high sugar cakes and puddings. Can you suggest lower-sugar cakes and lower-fat biscuits?

8 Working with some friends make a collection of recipes which use Fruits and Vegetables and unrefined Fillers, have low fat and sugar content and will please a particular group of people who interest you.

Part 2 Home Studies

8 Introduction: how do we furnish our homes?

Fig. 39

From the outside our homes often look almost the same as many others. Yet inside they are very different.

The people you see outside their homes in the picture have

bought different things for their homes, or arranged similar things differently.

In this section of Level 2 we are going to concentrate on the things we have in our homes. The things you have in your own home will be so familiar that you will find it hard to look at them with 'new' eyes.

How do we furnish our homes to suit our needs?

The things we have in our homes are there mostly because we want to use them. They have a job to do – or we once thought they had. We can get so used to familiar things that we put up with them when they are wearing out and no longer doing their job properly. We can become so envious of people with new things that we refuse to look around to see if we could use something we already have in a new way. But we can get pleasure out of trying to make our homes suit our needs.

 Find out . . .

. . . all the uses of the various things in your bedroom, apart from your clothes.

● At first glance some things will have only one use, but think again! Try to complete this table:

OBJECT	USES
Bed	For sleeping on ; for storing things under. For sitting on during the day ?
Wardrobe	For storing things ; for supporting a long mirror ?
Curtains	To cut out light and draughts ; to keep in heat ; to make the room look nice.
Bedcovers	
Carpet	
Wallpaper	
Chair	
Dressing table	

● How many things did you find in your room which were used for STORAGE? COVERING? SITTING ON?

One problem today may be lack of space in our homes. Manufacturers know this and so they set out to make DUAL-PURPOSE furniture. A bed becomes a settee during the day. A dressing-table becomes a desk when you want to do homework in peace.

Most people enjoy making their home just as they want them. Most of us have to buy mass-produced things but we can use them in interesting ways if we look at them with 'new' eyes.

● Work with a small group ● Make a display of pictures or drawings or, perhaps models, showing a variety of uses for different things people use around the home ● The illustrations in Fig. 40 are examples which will help you ● The group should look at things used for COVERING.

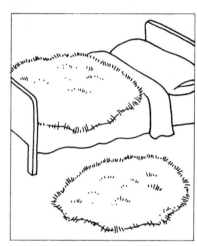

Fig. 40

● You will have to think about features such as AREA and THICKNESS, as you look for your new uses.

Another group should look at things used for DISPLAYING (we put things on them).

You have been looking at how we use ready-made things in our homes. Now we need to look at the materials used to make those things. Do materials have special uses or is one material sometimes used in different ways?

Find out ...

... how materials are used in the home.

● Make a list in your book of the ways in which these groups of materials are used in your home:

    (a) wood   (d) metals
    (b) plastics   (e) china and pottery
    (c) glass

● Try to find reasons why you think each material is used in the ways you have listed.
● Write sentences beginning like this:
Wood is used for tables and shelves because it is strong ....
● Compare your ideas with the rest of your group.

Practical activity

How many uses can you think of for wall tiles?

● Work with a small group to make an advertisement which will persuade people that wall tiles are just what they need ● You will need to think of all the things wall tiles can do. The pictures in Fig. 41 will give you some clues:

 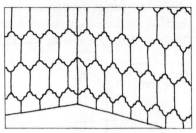 

Fig. 41

- Find words to describe tiles, for example:

  tiles are smooth; tiles can be pretty.

- Find words to describe what tiles can do, for example:

  tiles can cover surfaces; tiles can improve insulation.

- Then try to think of different uses for tiles apart from covering walls ● Remember that tiles come in different sizes and colours ● You can suggest uses for just one tile or for lots ● Use your imaginations and come up with lots of new ideas.

Discussion: are fabrics and carpets MATERIALS?

Fabrics are often called 'materials' but is this really right? Wood is a material. There are different kinds of wood but each kind is always wood. Look at a fabric and a rug and decide if they are materials. Use a lens and you will see that fabrics are made of fine fibres which may be made from different materials. Fabrics may contain, for example, glass or metallic fibres.

This is nylon fabric with fine fibres.

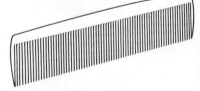

This is a plastic comb. It is made of nylon.

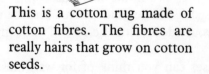

This is a cotton rug made of cotton fibres. The fibres are really hairs that grow on cotton seeds.

This is paper. It is made of mashed-up cotton rags and mashed-up wood.

Fig. 42

- Can you see that fabrics and carpets are NOT materials in the sense that wood is a material?

9 How many kinds of materials are there in your home?

 **Find out** . . .

. . . if you can name the materials these things could be made of:

(a) table fork
(b) cup
(c) washing-up bowl
(d) kitchen floor covering
(e) tumbler
(f) chest of drawers
(g) carpet with wool mark
(h) kettle
(i) window
(j) mixing bowl
(k) kitchen work surface

Discussion

Discuss your list of names with your teacher. Did you use group names such as wood or metal? Those are the names we have been using so far. Or did you use special names like stainless steel (metal), or oak (wood)?

If we sort the materials into groups or sets we can think about them more easily. All the materials in any one group have somethings in common. We have already sorted foods into groups so you will remember we used a special name. We CLASSIFIED foods. Here is one way of classifying materials:

METALS

aluminium (pan)
stainless steel (knife)
brass (candlestick)
silver (box)

WOODY MATERIALS

pine (table)
oak (door)
teak (board)
beech (wooden spoon)
paper (tissues)

STONY MATERIALS

china (cup)
glass (tumbler)
earthenware (basin)
stone (steps)

PLASTIC MATERIALS AND RUBBER

polythene (bag)
polystyrene (beaker)
nylon (pan cleaner)

TEXTILES

cotton (towel)
linen (tea towel)
wool (blanket)
nylon (velvet curtains)

Look at your list of names and see if you have materials from each group.

You can now classify most household materials under the five headings. You will have seen that nylon occurs in two groups and that there is a similarity between cotton and paper. This is because it is difficult to find a classification which has no overlaps.

Perhaps you are not sure of all the special names for materials in each group. There are too many of them to do in one go, so start with the metals and the metal mixtures, called **ALLOYS**.

Find out ...

... how many names of metals you know.

• See if you can match these objects with the name of a metal or alloy they can be made of – at least on the outside (for example, silver plating is only the outside coat).

(a) tap	(1) tin
(b) tubular chair	(2) copper
(c) hot water pipe	(3) silver
(d) candlestick	(4) aluminium
(e) saucepan	(5) chromium
(f) casserole	(6) steel (alloy)
(g) can of food	(7) iron
(h) chain around a girl's neck	(8) brass (alloy)

It takes a long time to recognise all the different materials. Sometimes one material is covered with another. Sometimes they do not look the same as we expect that material to look. Manufacturers keep inventing new things to do with materials in the factories. Technologists invent new materials. We need to keep up-to-date so that we know what we are buying for our homes.

Practical activity

Start to make a collection of labels and instruction leaflets from new things • Mount the labels on cards labelled with the group name of the material and the specific name.

STONY MATERIALS
STOVEWARE: GLASS.

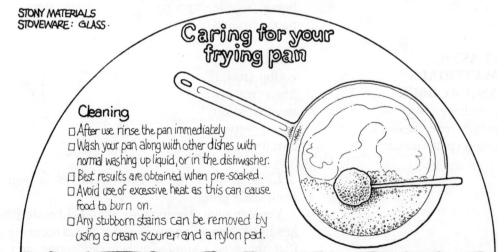

Caring for your frying pan

Cleaning
☐ After use rinse the pan immediately
☐ Wash your pan along with other dishes with normal washing up liquid, or in the dishwasher.
☐ Best results are obtained when pre-soaked.
☐ Avoid use of excessive heat as this can cause food to burn on.
☐ Any stubborn stains can be removed by using a cream scourer and a nylon pad.

Fig. 43

Why do we need to know the names of materials?

Discussion
How do these two materials behave:

WOOD   FIBRE GLASS

Fig. 44

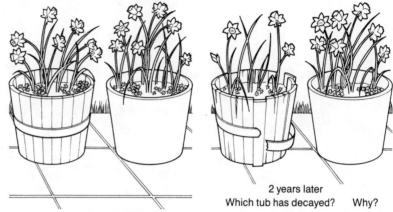

2 years later
Which tub has decayed?   Why?

We often have a choice of two or more materials which can be used to make the same sort of object.

Find out ...

... how many spoons made of different materials you have in your Home Economics room.

● Make a collection of different spoons – e.g. metal, wood, plastic.
● Label each with the group name of the material and with the name of the specific material it is made of. (Your teacher will help with those you do not know.)
● Try to describe them by asking these questions:

(a) What does it look like? (Colour? Shiny or dull?).
(b) How does it feel? (Cold if the weather is cold? Rough or smooth?).
(c) What happens if you gently try to bend it?
(d) What happens if you drop it?
(e) What happens if you put one end into boiling water?
(f) What happens if you bang it against metal?

● Write out a new label for each spoon describing how it behaves. Your new label will give the **PROPERTIES** of the material the spoon is made of. The spoon has its own properties because of the material from which it was made. 'Properties' is one of those words which has more than one meaning.

Discussion

1 Discuss with the rest of the class the PROPERTIES you have tried to describe. Who found the best words to use?
2 Which spoon could you leave in hot soup while it was cooking?

59

3 Check the properties you noted and see if they include these:

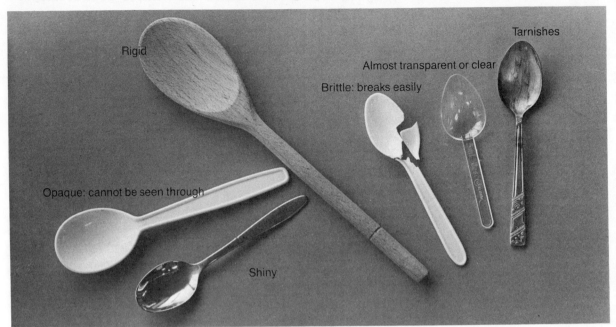

Fig. 45

4 Add these words to your vocabulary list and keep an eye open for new words which describe the properties of materials.

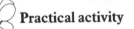 **Practical activity**

Choosing materials.

Imagine that you want to buy a beaker or mug and that your local shops sell only these:

Fig. 46

Before you decide which to buy:

- Decide why you need a mug – for a picnic? for flowers?
- Search for information • Find out prices • Find out which specific material each mug is made of • Look up the properties of each material.
- Choose the material you think will be best for your mug and say why.

10 Why do we choose different materials?

Find out . . .

. . . when you and a partner would use each item in the pairs listed below.

- Collect all the items ● Look at each item first and feel it ● Think of words to describe it.
- Explain to your partner WHEN you would use the first item in a pair ● Say WHY you would use it ● See if your partner agrees ● Let your partner do the same for the other item in the pair.

1 A GLASS tumbler and a POTTERY beaker.
2 A PLASTIC basin and a CHINA basin.
3 A LINEN tea towel and some PLASTIC food film.
4 A WOODEN chair and a METAL chair.
5 A STEEL table knife and a PLASTIC table knife.

- Did you and your partner agree on all items or just some? ● If you did not agree, why was that? ● Check your reasons with this background information.

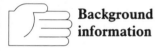

Background information

There are three main reasons why we think about materials for our homes.

1 The material will wear well. We say it is FIT FOR THE PURPOSE.
2 We like it. We like the way it looks and the way it feels. We say it has APPEAL.
3 We can afford it. The PRICE is right for us.

If you look at these reasons you will see that each of us has to decide about APPEAL and PRICE for ourselves. We can learn about **FITNESS FOR PURPOSE**.

We have to go on learning about materials all our lives, because they keep changing. New materials are made, and the old ones changed and made better. We all need to be able to find information which will help us to choose the best material for the job we want it to do in our homes.

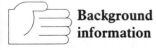

Background information

Finding out about materials.

We can look up information in books. We need to check when the book was published. We may not find anything about new materials, or new treatments for old materials, if the book is out-of-date.

We can ask other people how they have found the materials behave in use. We must make sure that we want to use it in the same way.

We can write to experts or listen to them on the radio or TV.

We can read advertisements and then check what they say. They may not tell us ALL we need to know.

We have to be able to ask the right questions and understand the answers. We have to be able to use the right words.

 **Find out** ...

... if you can understand the words underlined in these advertisements:

1 Kitchen paper
ABSORBS spills
instantly!

2 Glazed curtains with luxury SHEEN.

3 Carpet: really
DURABLE.

4 Hairdryer: safe, fully-INSULATED, with plastic cover.

Discussion

Discuss the meaning of the words with your teacher.

1 What do they tell us about the items?
2 Decide on a list of questions we need to ask about materials before we choose things for our homes. Use these clues to help you:
  (a) What can it do? (Paper ABSORBS spills.)
  (b) How does it look? (Glazed cotton has a shine.)
  (c) How does it feel? (Wool feels warm or, rather, not cold; metal often feels cold.)
  (d) How long will it last? (Woollen carpets are DURABLE.)
  (e) Will it be electrically safe? (Plastics are electrical IN-SULATORS.)
  (f) How much does it cost?
  (g) Do I like it?

 Practical activity

Collect labels showing symbols which give us information about materials.

- Do the labels tell you what you want to know?
- Are the labels clear so that you understand what they are telling you?
- Labelling is still fairly new • Look out for new labels and improvements to old ones.

How have things we use in the home developed from the past?

You may have dressed up in a historical costume for a school play, or you may have visited a museum and seen a collection of toys. Perhaps you have seen a TV programme about antiques or something may have caught your eye in a local antique shop window. If you have done any of these things you already have some ideas about how things have changed over time.

Finding out how things have developed from the past is a hobby for some people.

Find out . . .

. . . some of the reasons why things we use in the home have changed over time.

- Try to answer the questions under each set of pictures in Fig. 47. They are all about containers for holding liquids.

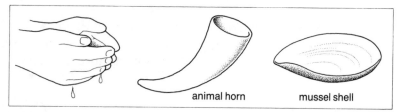

animal horn   mussel shell

- What did our earliest ancestors probably use to hold water?

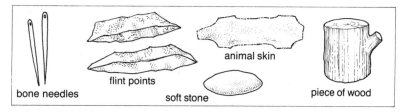

bone needles   flint points   animal skin   soft stone   piece of wood

- What might they have used to make their own containers?

Bronze Age Baked Clay Pots

- How did man make better containers?

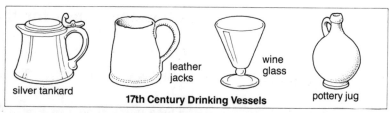

silver tankard   leather jacks   wine glass   pottery jug

17th Century Drinking Vessels

- How did we get a greater variety to choose from?

63

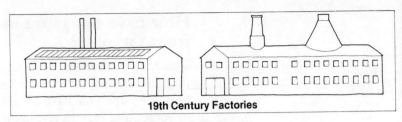

19th Century Factories

● How did we make larger quantities?

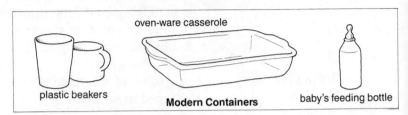

oven-ware casserole

plastic beakers   **Modern Containers**   baby's feeding bottle

Fig. 47

● How and why do we keep on inventing new materials and improving old ones?

Practical activity

Try to visit a museum ● Look particularly at how one kind of item we use at home has developed over time. If you look at chairs for instance you may see some of these:

Fig. 48

Choosing things for the home

Questions

1 WHEN IS A CUP A GLASS?

Extra work

These items shown in Fig. 49 are all used to hold liquids or to drink from.

1 Wine glass

2 Beer jug

3 China cup and saucer

4 Chinese tea cup

5 Pewter tankard

6 Sherry glass

7 Invalid feeding cup

8 Bowl and spoon

9 Breakfast bowl

10 Tea glass with metal handle

11 Wooden bowl

12 Bottle

13 Plastic tumbler

14 Glass tumbler

15 Plastic mug

16 Enamel cup

Polystyrene beaker

Fig. 49

(a) Classify the materials used under the same headings as before.

(b) Choose THREE containers suitable for the following purposes. Give your reasons.

   (i) For a young child just learning to drink without help.
   (ii) For drinking soup.
   (iii) For drinking lemonade on a picnic.

(c) Look at the problems and check your choices.

   (i) What would be the problems of drinking tea in container 14?
   (ii) Why do people not usually drink wine in container 5?
   (iii) What would be the problem if you wanted to drink from 4 or 9 and they had been filled to the very top?

(d) Consider what people like. Which containers appeal to you? Why?

2 DECIDE YOUR CRITERIA FOR CHOOSING DIFFERENT ITEMS.

This burr shows one pupil's criteria for a cup.

Fig. 50

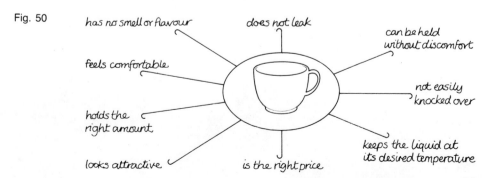

has no smell or flavour
does not leak
can be held without discomfort
feels comfortable
not easily knocked over
holds the right amount
keeps the liquid at its desired temperature
looks attractive
is the right price

65

3 CHOOSE DRINKING VESSELS FOR THESE PEOPLE:

(a) A couple is getting married. They already have some everyday cups but think they would like something special for 'best'. Their parents want to buy them a present. What should they buy? Any of these? Why?

Range of 'Cups'
- Bone china cup and saucer – £200 for 6.
- Decorative mugs (pottery) – £1 each.
- Tea set of everyday china – £25 for 6 cups and saucers.
- Glass beaker with metal handles – £2 each.
- Aluminium mugs with handles – £1 each.
- Decorative plastic mug in holder – 90p each.

(b) Paul is moving into his first flat. He needs a set of 'cups' but is short of money. Which might he choose? Why?
(c) The Smiths take packed lunches. They like soup in the winter, but this means taking drinking containers each day. Which might they choose? Why?
(d) Amany likes black coffee and tea. Her family keep to their traditional Middle Eastern ways. What kind of 'cups' might she choose? Why?

4 CHOOSING AN ITEM OF YOUR CHOICE

(a) Choose any item we use at home.
(b) Define its purpose – say what it is normally used for.
(c) Describe and record with pictures if possible the range of choices available – what can you choose from in the shops?
(d) Note these things:

   (i) the different materials used;
   (ii) any different shapes/colours/patterns etc;
   (iii) any extra uses they have;
   (iv) the different prices.

(e) Make a choice check-list for this item (i.e. decide your criteria).

11 How can we try to keep things looking as good as new?

If you have recently had something new which you really wanted you will remember how you felt and behaved at the time. Perhaps others in your class have had similar experiences.

 **Find out** ...

... the way we feel about new things.

- Work in a small group, and share experiences.
- Tell each other how you felt and what you did when you got something new. Try to choose something you got a little time ago. Say about a month ago.
- Note any words which most of you used – perhaps most of you said you were EXCITED.
- Note any actions which most of you said you did – perhaps you said you kept TOUCHING your new possession.
- Go on to discuss how you felt and behaved towards your possession after some time and now. Have you changed your feelings towards your possession? Why?
- Did other people have similar feelings to yours?
- Did they behave in a similar way?

 **Practical activity**

Use your understanding of how people feel about new things to write a short story about what happened to a new pair of boots.

The pictures in Fig. 51 show you what happened to a beautiful pair of new boots. You can invent who owned them and why that person wanted them so much. You must also invent what happened to the boots in the end. Write the full story in your own words.

Fig. 51

   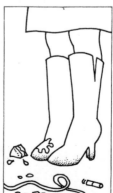 

After writing your short story you can see that one reason why our things do not stay looking like new is that they get damaged. If they suffered less damage they would last longer. So we need to look at DAMAGE more closely. In your story the boots were damaged by rain and by a sticky piece of cake.

Discussion

1 Discuss accidental damage to our things.
2 Work in a small group and share your experiences.
3 Think back to when one of your possessions was accidentally damaged.
4 Try to remember just what happened. Then work out why things happened.
5 Make a note of any words or groups of words which most of you used (as you did when you were thinking about new things).
6 See if your ideas explain the situations shown in Fig. 52:

Fig. 52

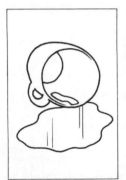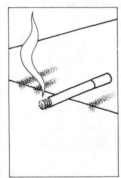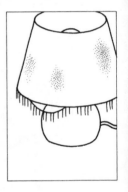

 Practical activity   (a) Choose one of the pictures above and try to answer these questions about it in your book.

   (i) Why did the person who caused the damage do what they did?
   (ii) What should they do now?

(b) Work out a way of recording damage to objects in the Home Economics room.

   ● Try to classify:
   (i) the kind of damage (cuts, breakages, etc.);
   (ii) the causes of the damage (clumsiness, ignorance, etc.).

   ● Keep a record for a convenient length of time and then discuss your findings.

   ● Try to decide:
   (i) how we can reduce damage by THINKING in advance;
   (ii) how we can reduce damage by what we DO to our possessions.

In the next section we shall look at some of the things we can do to our possessions to keep them looking like new as long as possible. One of the simplest things we can do is to keep things CLEAN. Dirt spoils the look of things. It can also damage them.

What is dirt?

Discusssion

Discuss and record what makes things dirty.

1 Work in groups and make a list of all the things that have to be cleaned in a kitchen, and the kind of material they are made of.
2 Describe what the dirt is like. These words will give you clues:

GREASY, DUSTY, STAINS, DRIED FOOD.

Sometimes it may be a mixture of different kinds of dirt.

3 Look at these pictures and discuss what kinds of dirt would get into the homes in Figs. 53 and 54.

Fig. 53

Fig. 54

 **Practical activity**

● Make samples to represent THREE kinds of dirt.

1 Any fine dry powder such as chalk dust.
2 Some coarse powder such as sand, just moistened with water.
3 Sand moistened with any grease or heavy oil.

● Smear each kind of dirt on to pieces of card covered with foil shiny side up ● Leave the dirt to settle for a few minutes.

69

- Record the appearance of the foil when you try to remove each kind of dirt in the following ways.
    (a) Using a soft dry duster.
    (b) Using a moist soft cloth.
    (c) Using a soft cloth moistened with warm detergent solution.

- Try to explain your findings.

What kinds of cleaning products do we use in the home?

If you think back to the work you did on classifying kinds of dirt you will see that we need things which will remove:

(a) dry dust;
(b) moist dirt;
(c) greasy dirt;
(d) stubborn dirt (stains);
(e) other problems such as a change in the outer layer of the material. This can build up on the surface of metals. We call it **TARNISH**.

You also know that cleaning products have to remove dirt from different surfaces without spoiling or damaging them.

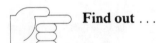 **Find out** ...

... how you would remove dry dust from these surfaces.

Surfaces
(a) Rug   (e) Hard floor
(b) Carpet   (f) Curtain
(c) Table   (g) Mirror
(d) High ledge   (h) Radiator

- See if you can match the aids or equipment in Fig. 55 with each surface ● One aid or one piece of equipment may do more than one job.

Fig. 55

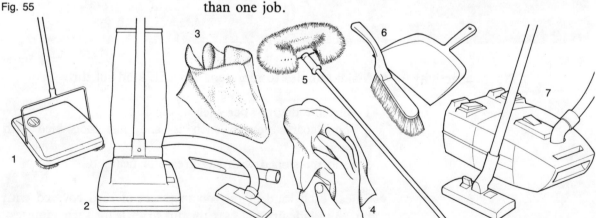

Discuss your choices with a partner and note similarities and differences. Check your findings with this statement:

DRY DUST MAY BE REMOVED BY:

    (a) PICKING IT UP
    (b) SUCKING IT UP
    (c) BLOWING IT AWAY

Did you choose all of these methods? Discuss their advantages and disadvantages.

Cleaning in the home

Having got rid of dry dust we are left with all the other kinds of dirt. For removing these we use different cleaning products – such as powders, creams and liquids. If a surface or material is not damaged by water, we can remove most kinds of dirt by different ways of washing. The simplest form of washing is wiping with a damp cloth.

Find out . . .

. . . the best way to wipe a plastic surface.

- Decide how you want the surface to look when you have finished. For example, is it to be shiny? Is it to be free from smears?
- You are going to try to test different results using only a cloth and water.
- Decide: (a) which types of cloth you will compare;
  - (b) the different temperatures of water you will try;
  - (c) the different movements you will compare for the wipe;
  - (d) any other tests you think are important.
- Work in groups so that you do not run out of surfaces in the Home Economics room.
- Plan fair tests and record your findings.
- Try to account for any problems.

Practical activity

Try to match these problems about wiping with the solutions.

A FAMILY HAS HAD DINNER AT A PLASTIC-TOPPED TABLE WITHOUT A CLOTH. THERE ARE CRUMBS AND SPILLED, DRIED STEW ON THE TABLE.

Problems
(a) How can you avoid crumbs on the floor?
(b) How can you avoid leaving smears on the table?

(c) How can you avoid leaving bits of hard food on the table?

(d) How can you avoid leaving the table wet?

Solutions

1 Wring out the cloth really well for the final wipe.

2 Keep using new parts of the cloth.

3 Rinse the cloth.

4 Pick up with the cloth – fold cloth and cover, rather than wipe.

5 Use a different method. Wiping isn't good enough!

When wiping is not good enough we usually try to wash dirt away. The products we use for washing all contain **DETERGENTS** and other ingredients which help the Detergents to dissolve some soluble dirt and remove greasy dirt from surfaces.

 Find out . . .

. . . more about washing-up.

You should be able to work out your own fair tests. Here are some ideas:

WASHING-UP TESTS

1 Compare milky tumblers washed in (a) plain water and (b) Detergent solution at the SAME TEMPERATURE.

2 Compare greasy plates washed in different concentrations of Detergent solution at the SAME TEMPERATURE.

3 Compare greasy plates washed in (a) cold and (b) hot Detergent solution, using the same CONCENTRATION of Detergent solution.

4 Compare the number of plates washed before the surface bubbles vanish using the SAME AMOUNTS of two DIFFERENT DETERGENTS at the SAME TEMPERATURE.

 Practical activity

Make a poster about washing-up.

● Use the findings from your tests to remind others how to wash-up quickly and well ● Here are some jumbled points for you to consider:

RINSING	KIND OF DETERGENT	PREPARATION
TEMPERATURE	METHOD OF WASHING	DELICATE ITEMS
AMOUNT OF DETERGENT	ORDER OF WORK	SPECIAL PROBLEMS

 Extra work

1 Investigate dish-washing machines.

If you cannot use an actual machine collect information about:

(a) the kinds of things they can wash;

(b) the way they have to be loaded;

(c) the temperature at which they wash;

(d) the movement of the water;

(e) the rinsing of the dishes;

(f) the washing and rinsing times.

● Discuss the advantages and disadvantages of washing-up machines for different people ● Why are they used in places where large numbers of people eat?

2 Investigate what fabric-washing machines do.

● Look at a machine in your Home Economics room if you have not already used it and note the controls, as you did for the cookers ● Note any similarities and differences between dish-washing and fabric-washing machines.

Fabric-washing machines control most of the washing process. We have to start them, select the wash programme, and put in the kind and amount of Detergent or powder we think is right.

So, your next investigation must be into KINDS of fabric-washing powders.

How do we use washing powders with fabrics?

 **Find out** ...

... first about the different KINDS of washing powders we can use.

● Collect as many packets and containers of these as you can find in the Home Economics room. Each will have a different name, but the name will not tell you what kind of mixture is in the pack. You may be able to guess by looking at the pictures, but you will get a fuller idea if you make a table like the following one. (The list on the left shows some of the things a family might want to wash, and their fabric names shown on the care labels.)

(a)	Washing Powders			
	W	X	Y	Z
woollens				
white cotton shirt				
blue cotton shirt				
polyester/cotton sheets				
acrylic dress				
lace blouse				
white nylon blouse				
terylene curtains				
coloured cotton towels				

● You put a tick for each item included in the instructions on the pack.

● The powders which have ticks in the same places are of the same KIND.

● You should also make another table like this one for wash codes. (In Level 1 you saw that the number underneath the 'water level' is the temperature that the washing water should be.)

(b) Wash Codes	Washing Powders			
	W	X	Y	Z
A				
B				
C				
D				

● You put a tick for each code included in the instructions.

● The powders which have ticks in the same places are of the same KIND.

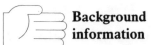

Background information

Washing powders are designed to do these things:

1 Help to remove dirt safely from certain fabrics. (Without damaging the fabric.)
2 To work best at the temperatures which are safe for those fabrics.
3 To help to remove the kinds of stains likely to be found on those fabrics.

Practical activity

Investigating the important points about fabric washing powders.

● Plan and carry out tests like these:

1 Compare the results of washing woollen fabric in (a) warm water and (b) very hot water. Here are some clues to guide you so that your comparison is fair:

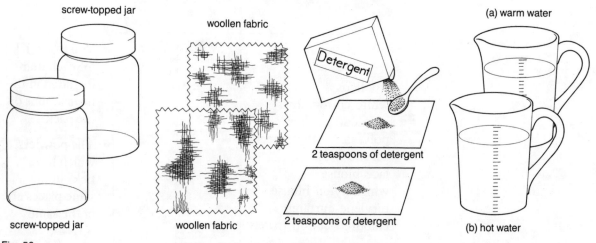

screw-topped jar

woollen fabric

(a) warm water

2 teaspoons of detergent

screw-topped jar

woollen fabric

2 teaspoons of detergent

(b) hot water

Fig. 56

For quick results the hot water should be as hot as your teacher thinks is safe.

• Shake the two jars equally for 5 minutes, then rinse the pieces of fabric in warm water and dry them side by side away from direct heat • When they are dry, note any differences in appearance, feel and size. Not all woollen fabrics behave in exactly the same way, so you may find that you have to repeat the test on another fabric before you get very marked differences.

2 Compare the results of gentle washing with harsh washing on woollen fabric.

• Wash two similar pieces of woollen fabric in the same warm solution for the same length of time, but by different methods:

    (a) squeeze one piece gently without twisting it;

    (b) rub the second piece between the hands so that the surface of half rubs against the surface of the other half.

3 Try other comparisons such as these:

    (a) Washing very dirty cotton fabric with and without pre-soaking.

    (b) Washing non-fast coloured fabric in cold and hot water. (Yellow duster fabric is a good example.)

These tests will have shown you how much you could learn by studying more about fabrics in a special textile study course.

How many ways are there to remove dirt from hard smooth surfaces?

Discussion
How would you remove dirt in these cases? Why?

Fig. 57

Washing is sometimes not strong enough. We may have to rub the surface clean. Sometimes we also have to use what is called an **ABRASIVE CLEANER**. An abrasive cleaner is used to rub down a surface so that dirt or paint is worn away.

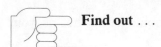

Find out . . .   . . . about abrasive cleaners.

● Find as many examples as you can see in your Home Economics room. You will probably find most of these:

(a) scouring pads – round or flat in different materials
(b) scouring powders
(c) scouring pastes
(d) scouring liquids
(e) scouring brushes
(f) cleaning cloths (you do the rubbing!)

● Work in two's ● Look closely at one abrasive cleaner
● Find words to describe its feel, smell, shape, colour and how you think it will remove dirt.
● Report back to the rest of the class.

Practical activity   Testing methods of cleaning surfaces.

● Work in two's and prepare a report.
● Carry out a fair test and recommend a suitable abrasive cleaner for a surface in your Home Economics room. Add your recommendations to the care label.

This is what one school did. They chose aluminium foil as an example of a polished surface.
1 They took a piece of card and covered it in foil, shiny side up.
2 They taped areas for the test like this:

1	2	3	4	5	6	7	8	9

3 They rubbed each section with a different abrasive cleaner from this list:

(a) green pad	(e) abrasive powder
(b) steel wool	(f) nylon scourer
(c) liquid cleaner	(g) disposable cloth
(d) brush	(h) dish cloth
	(i) silver polish

4 The same pupil in each pair rubbed the foil for each test.
5 Each section was rubbed FOUR times, equally hard (fair test).
6 They noted the effects and then ranked the 'cleaners' starting with the one which had done the least damage to the foil and ending with the one which had done the most damage. Their recommendation was 'Use a disposable cloth or a soft brush to remove sticky dirt from polished aluminium'. They added another recommendation based on their own experience: 'Some metals tarnish. A silver spoon tarnishes if used for a

boiled egg. Tarnish has to be removed with special chemical cleaners'.

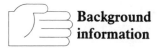

Background information

Cleaners

(a) Cleaning powders and pastes often contain very finely ground mineral powders which abrade stubborn dirt.

Mineral powders are graded and tested on a range of surfaces. Soft surfaces are scratched by coarse hard powders.

Manufacturers say that the powder which is included in a cleaner is always one grade finer than the grades which could scratch the kind of surface it is meant to clean.

(b) Cleaning cloths and pads abrade because this surfaces are sharp or coarse. They vary in their coarseness.

Questions

1 Which abrasive cleaner would you use on a relatively soft material?

Name a surface in your home made of such material.

2 Name a relatively hard surface in your home. What would you use to remove stubborn dirt from that surface?

3 Write sentences to show that you know how to use the words: **ABRASIVE, ABRASION** and **ABRADE.**

Extra work

People used to have to spend a lot more time cleaning things than we do now. Talk to older people and find out what they had to clean and what they used to do it. Try to collect old recipes for scouring powders and metal polishes.

Part 3
Reference Section

Microscopes

A microscope magnifies things. It helps you to see some things more clearly. It also helps you to see things you could not see otherwise. Like the telescope, the microscope was a great help to scientists when it was invented. It was a scientific breakthrough. Modern microscopes are a great improvement on earlier types.

Construction of a microscope

The microscope in your Home Economics room may look different from the one you have used in Science classes. Look carefully at the Home Economics microscope and see if you can identify the main parts. All microscopes work on the same principle, so they must have the same basic parts.

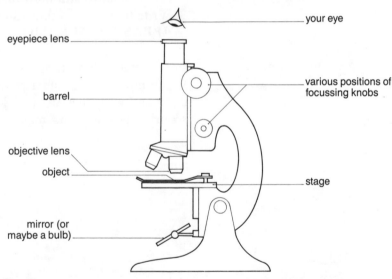

Fig. 58

How they work

You see the object magnified by two sets of lenses. The **OBJECTIVE** lens is near to the object. The **EYEPIECE** lens is nearest to your eye. You get the magnification you want by choosing different objectives or eyepieces. They have numbers engraved on them. If the eyepiece has '×6' and the objective '×4' your object will be magnified 24 times (6 × 4 = 24).

Always start on a low magnification. Use only very low power (×4) or low power (×10) objectives.

Lighting the object

The object can be lit in two ways:

(1) Top lighting which works like this:

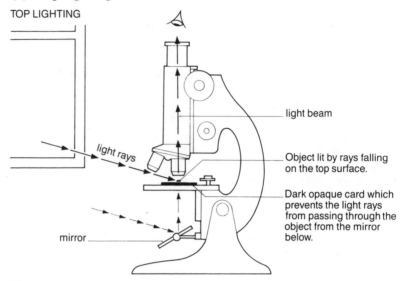

TOP LIGHTING

light beam

light rays

Object lit by rays falling on the top surface.

Dark opaque card which prevents the light rays from passing through the object from the mirror below.

mirror

Fig. 59

(2) If the object is thin enough, light can shine through it. Many objects are semi-transparent if you use thin layers. You see a picture of the inside not just the outside. THROUGH lighting works like this:

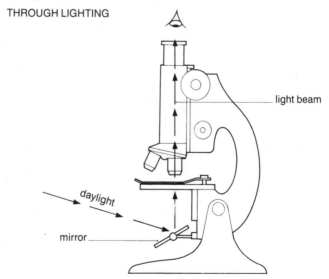

THROUGH LIGHTING

light beam

daylight

mirror

Fig. 60

Ask your teacher if you may practise looking at objects using both top and through lighting. You could look at fabrics, sugar or salt crystals.

79

You can buy specially prepared **SLIDES** of some objects from dealers. Different dyes are used to show up different parts. But slides are expensive, and it is more interesting to make your own.

Looking at objects

1 To get practice in using top lighting you can look at the scratches that were made when you tested the cleaning powders on aluminium foil. You could also look at a salt crystal.

    (a) Put a crystal of salt on to a square of black card or paper.

    (b) Judge when you think the crystal is in line with the barrel of the microscope by looking from the side.

    (c) Using the lowest magnification possible, look down the barrel and find the crystal.

    (d) You will need to focus the microscope. This will be easier if you begin by having the barrel as low down as you can. Then when you focus you will only have to turn the focussing knob slowly in one direction.

    (e) Draw what you see.

Fig. 61   Microscopic view of salt crystals.

To get practice look at the parts of an unpolished brown rice grain. Compare it with a polished grain.

2 To get practice in using through lighting you can look at vegetable cells, amongst other things. This time you will need clean glass slides. Onion cells are good to look at because you can peel off a layer of membrane which is only one cell thick.

    (a) Tear off an onion 'leaf' carefully, hoping to leave rough edges of this transparent, flimsy membrane.

Cover slips are very thin pieces of glass which shatter very easily. Never use them in a kitchen. They are dangerous because pieces can get into food without you noticing.

(b) Carefully cut off a small piece and place it in a drop of water on a slide. The piece of onion should be as flat as possible.

(c) Using very low power magification focus on to the membrane. If a cover is needed to flatten the membrane use another clean slide.

Air can get into home-made slides and this is a problem. You need to know how to tell air bubbles and air pockets from cells. Compare this view of onion cells with the view on page 8. Air bubbles have bright hard edges. The bubbles really stand out from the background.

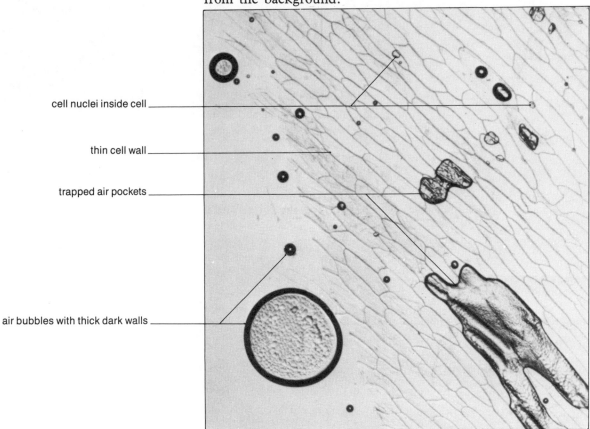

cell nuclei inside cell

thin cell wall

trapped air pockets

air bubbles with thick dark walls

Fig. 62   Microscopic view of onion cells showing air bubbles on slide.

3 To get practice look at other foods under the microscope. These points will help you.

(a) To start with always use cooked foods.
(b) Sometimes the food may be too thick. Make it thin enough by squashing it between two slides.
(c) If you make a poor slide do not spend a long time trying to get it right. Make another one.
(d) If you only see a dark mass, you have probably got too much on the slide. You really need very little.

Using prepared slides
Figure 63 is a photograph of a slide showing what raw potato cells look like. The colours are named because this book is in black and white only.

loose starch granule stained dark blue

cell wall stained brown

empty cell

starch granules still inside the cells

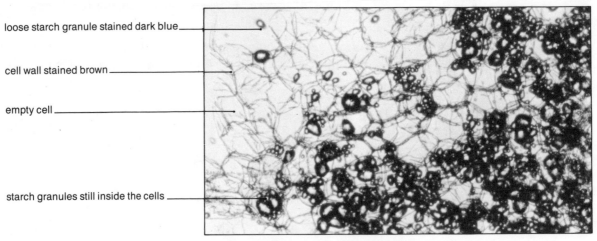

Fig. 63   Microscopic view of raw potato cells.

Compare this with an easily made home-made slide of cooked potato. Put a little 'pin head' of cooked potato on to a drop of water on a clean glass slide. The potato is **DISPERSED** or spreads out on the slide. You may need to put a second glass slide on top.

cells have rounded off in shape

separated but whole potato cells

gelatinised starch granules inside the cell

a few cells broken when the slide was made

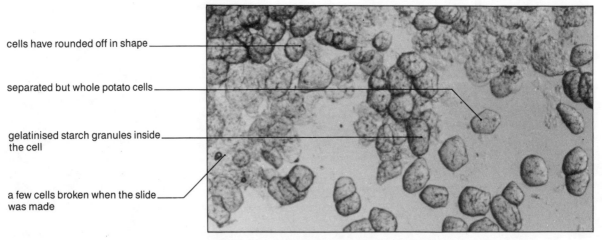

Fig. 64   Microscopic view of cooked potato cells.

Colouring slides to look at starch granules
Start by looking at potato starch granules, because they are larger than the starch granules in other foods.

(a) Scrape off a little raw potato with the edge of a knife.
(b) Put this watery scraping on to a slide.
(c) Add a drop of weak iodine solution to colour the starch granules.

82

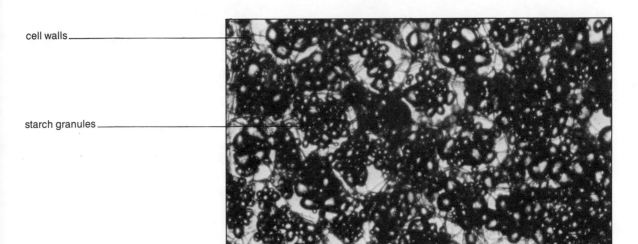

cell walls

starch granules

Fig. 65 Microscopic view of stained raw potato cells.

After looking at slides of raw and cooked potato cells and after seeing the starch granules clearly you should have a better understanding of what happens when a potato is cooked.

There are many types of starch granules mentioned in this book. The other types may be smaller than potato starch granules but they behave more or less in the same way.

As you work with a microscope think back to work you have done on potatoes and grains. Add extra notes and comments to those you made at the time before you saw the slides.

Food labelling laws

The food labelling laws make sure that we get all the information we need when we buy food. The laws are designed to protect us. They can do this only if we make use of the information given on the labels. We have to make the effort to understand the labels. Everyone should read this booklet.

LOOK AT THE LABEL

A booklet by the Ministry of Agriculture, Fisheries and Food explaining what the labels on prepacked food mean, and why we need them.

Fig. 66

83

Customary names

If a food has always been known by a certain name we are quite clear what the name means. For example, a Chelsea Bun always looks the same or at least it should do!

Descriptive names

Foods without Legal or Customary names must be given Descriptive names. These must describe the foods accurately. For example, strawberry yoghurt must contain strawberries. But strawberry flavoured yoghurt need not contain strawberries.

Process name

The Process name must tell us the way the food has been prepared or processed. For example, *chopped* nuts or *powdered* nutmeg.

Ingredients

The ingredients of most pre-prepared foods must be listed on the packet or can.

The foods must be listed in order, starting with the one which is used in the largest amount.

In this example, water is the largest ingredient and spices are the smallest ingredient.

water
peas
bacon
ham
sugar
salt
starch
hydrolysed veg. protein
flavouring E150
spices

Special note: water is not always included in the list of ingredients.

Additives

Flavourings, preservatives, and colourings are listed under code names, e.g. E150. They can be used only if they have been coded by scientists who have tested them to make sure they are safe to eat. A complete list of additives can be obtained from the

Ministry of Agriculture, Fisheries and Food. The list is included in the booklet 'Look at the Label.'

Food Tables

Amount of sugar (to the nearest gram) in 100g of food

MAINS (M)	g in 100g	Average Portion (g)	No. of teaspoons of sugar
These are very low in sugar except for milk			
Cows Milk	5	–	–
Some MAINS have added sugar and are DOUBLES			
Cheesecake (M & F)	14	100	3
Fruit Yogurt (M & F)	18	160	6

FILLERS (F)			
White Bread	2	–	–
Wholemeal Bread	2	–	–
Boiled Potatoes	0	–	–
Boiled Sweet Potato	9	150	3
Boiled Rice	0	–	–
Lentils	2	–	–
Baked Beans	5	Small can	2
Processed Peas	1	½ med can	v. low
Roasted Peanuts	3	50g	v. low
Sultanas	65	–	–
Prunes (dried, raw)	40	5 fruits	4
White Sugar	100	–	20
Brown Sugar	99	–	19¾
Golden Syrup	79	2 tsps	1½
Honey	76	2 tsps	1½
Jam	69	2 tsps	1½
All Bran	15	25g	1
Porridge	0	–	–
Sugar Puffs	56	25g	3
Wheetabix	6	2 pieces	v. low
Currant buns	14	1 bun	1¼
Malt Loaf	19	1 slice	1
Low Sugar Scone	6	1 scone	v. low
Iced Fancy Cakes	54	1 cake	2¾
Ginger Nut Biscuits	36	1 biscuit	1
Sweet Pickle	33	3 tsps	¾
Butter	0	–	–
Ice Cream	23	Cornet	6
Apple Pie	31	100g	6
Rice Pudding	9	½ small can	2½
Fruit Flavoured Jelly (made with water)	14	¼ whole	4
Chocolate Bar	56	1 small (60g)	7
Fruit Gums	43	1 tube	2½
Liquorice Allsorts	67	Small pkt	14
Mars Bar	66	1 bar	9
Toffee	70	1 tube	7

FRUITS AND VEGETABLES (F & V)	g in 100g	Average Portion (g)	No. of teaspoons of sugar
Eating Apple	12	1 apple	2
Banana	16	1 banana	3
Lemon	3	–	–
White Grapes	16	50g	1½
Plums	10	3 plums	3
White Cabbage	4	100g	1
Garden Peas (Boiled)	1	80g	v. low
Carrots	4	100	1
Sweetcorn	2	80	½
Tomatoes	3	1 tomato	v. low

WATERY DRINKS (WD)			
These are mainly doubles			
Coffee (unsweetened)	0	–	–
Tea (unsweetened)	0	–	–
Canned Tomato Soup (F & V + WD)	3	½ can	1
Tomato Ketchup (WD + F)	23	3 tsp	1
Salad Cream (WD + F)	13	3 tsp	½
Coca-Cola	11	1 can	7
Orange Drink (undi-luted)	29	200ml	2½
Ribena (undiluted)	61	200ml	5½
Lucozade	9	200ml	3½
Low-Cal Drinks	0	1 can	0
Orange Juice (canned, unsweetened)	9	200ml	3½

1 tsp sugar weighs approx 5g

KEY: (–) not appropriate to give portion sizes.
Source of tables: adapted and modified from *The Composition of Foods*, 4th Edn, 1978, HMSO.

Amounts of fat and water (to the nearest gram) in 100g of food

MAINS (M)	FAT g	WATER g	FILLERS (F)	Fat g	Water g
Boiled Egg	11	75	Boiled Potatoes	0	80
Scrambled Egg (with butter)	23	62	Chipped Potatoes	11	47
Fresh Whole Milk (M & WD)	4	88	Potato Crisps	36	3
Fresh Semi-Skimmed Milk (M & WD)	2	90	Boiled Yams	0	73
Fresh Skimmed Milk (M & WD)	0	91	Boiled Spaghetti	0	72
Goats Milk (M & WD)	5	87	Wholemeal Bread	3	40
Evaporated Milk	9	67	Fried White Bread	37	4
Plain Yogurt (skimmed milk)	1	86	Wholemeal Roll	3	29
Cottage Cheese	4	79	Chapati	13	29
Curd Cheese	10	80	Baked Beans	1	74
Edam Cheese	23	44	Processed Peas (Canned)	0	72
Cheddar Cheese	34	37	Masur Dahl	3	78
Baked Cod	1	71	Roasted Peanuts	49	5
Cod Fried in Batter	10	61	Fresh Coconut	36	42
Fried Fish Fingers	13	56	Prunes (Dried)	0	23
Sardines in Tomato Sauce	12	65	Sultanas	0	18
Roast Chicken (meat and skin)	14	62	Low Fat Spread	41	57
Roast Chicken (meat only)	5	68	Soft Margarine	81	16
Stewed Liver	8	62	Butter	82	15
Grilled Rump Steak	12	59	Ghee	100	0
Stewed Beef Steak	11	57	Vegetable Oil	100	0
Stewed Mince	15	59	Mayonnaise	79	18
Grilled Lamb Chops	23	36	Double cream	48	49
Roast Breast of Lamb	37	44	Single Cream	21	72
Grilled Bacon	35	35	Ice Cream	8	66
Grilled Pork Chops	19	36	Scones	15	22
Roast Pork	20	52	Scotch Pancakes	12	41
Pork Sausages	25	45	Currant Bun	8	29
Beef Sausages	17	48	Doughnut	16	26
Frankfurters	25	60	Victoria Sponge	27	15
Fried Beefburger	17	53	Swiss Roll	5	25
Corned Beef	12	59	Custard Cream Biscuits	26	3
Luncheon Meat	27	52	Ryvita	2	6
Boiled Ham	5	73	Water Biscuits	13	5
Cornish Pasty (M & F)	20	39	Rice Pudding	4	72
Pork Pie (M & F)	27	37	Apple Pie	16	23
Sausage Roll (M & F)	36	23	Honey	0	23
			White Sugar	0	0
WATERY DRINKS (WD)			Jam	0	30
			Fruit Gums	0	12
Tea (no milk)	0	100	Toffee	17	5
Orange Juice	0	89	Mars Bar	19	9
Cream of Tomato Soup	3	84			
Salad Cream	27	53			

FRUITS AND VEGETABLES (F & V)

	FAT g	WATER g
Apple	0	84
Banana	0	71
Grape	0	79
Oranges	0	86
Fruit Salad (canned)	0	71
Olives	8.8	77
Avocado Pear	22	69
Onions	0	93
Green Cabbage	0	97
Carrots	0	92
Sweetcorn (canned)	1	73
Tomatoes	0	93

Index

The pages on which special terms are first introduced are shown in bold figures.

Abrasive/abrasion/abrade	**77**
Abrasive cleaners	
fair test for	76
kinds of	76
Absorbs	62
Absorption	**14**
Aerated/aeration	**38**
of bread dough	43
of short crust pastry	50
Batter	**41**
Botanical names	
of plants	2
Bread	
comparison of recipes	43
ingredients for	42
kinds of	42
role of ingredients	42
Bread dough	
aeration of	43
Carbon dioxide	**36**
Cereal grains	
kinds of	13
processing of	17
Cereals, different meanings of	13
Cleaning cloths and pads	
range of	76
Cleaning powders and pastes	
composition of	77
Cleaning products	
kinds of	76
Cornflour	
behaviour with water	20
making a thick sauce	21
microscope view of	20
Criteria	51
Damage, to things used in the home	68
Detergents, in washing products	**72**
Dirt	
kinds of	69
removal from hard, smooth surfaces	75
removing by washing	73
Dispersed	**82**
Dough	
addition of fat to	43
meaning of	43
Dry foods	
buying of	11
classification of	11
storage of	11
Dry fruits, making a pudding of	11
Dual-purpose	54
Durable	62
Dust, methods of removing dry dust	71
Extracted	**29**
Eye-piece, of a microscope	78
Fabric washing	
kinds of powders for	73
machines for	73
Fair test	**44**
Fat	
content in food portions	32
content in 100g of foods	87
meanings of to different people	26
Fats	
content of spreading fats	31
fatty Mains	31
hidden fats in foods	31
Fatty tissue, where found in the body	26
Fitness for purpose	**61**
Flour	
methods of making	19
types of	19
Flour mixtures	
behaviour when heated	35
kinds of	35
Food tables	
fat content	87
sugar content	86
water content	87
Fruit, plant cells	8
Gelatinisation	**22**
Ghee	30
Home	
choosing things for	65

furnishing to suit our needs 54
groups of materials used in the home 57
historical development of things used in the home 63

Insulated/insulators **62**

Materials
   classification of 57
   properties of 59
   reasons for choosing 61
Metals, names of 58
Microscopes
   construction of 78
   eye-piece lens for 78
   objective lens for 78
Muesli, making your own mix 19

Objective lens, for microscope 78
Oils 29
   extraction of 29

Pancakes
   comparison with bread and biscuits 41
   making of 37
Pastry
   consistency of short crust pastry dough 48
   ingredients for short crust pastry dough 50
   uses of 48
Plants
   botanical classification 2
   cell walls 8
   edible parts of 2
   Home Economics classification of 2
   sap 8
Polyunsaturated fatty acids **31**
Polyunsaturates **31**
Portion **24**
Processing 12
Properties of materials **59**
Protoplasm **28**
Pulses 12

Ratio 43
Refining **18**

Rice
   cooking of 13
   kinds of 14
   ways of boiling 14
Rubbed-in mixtures 46

Salads
   definition of 6
   names of 6
Savoury **15**
Scones
   cooking of 44
   cost of 47
   ingredients for 47
   investigating cooking of 44
   making of 47
   what are they? 44
Scotch pancakes
   appearance of 37
   choosing method of making for 38
   ingredients for 38
Soup, making vegetable soup 4
Sugar
   contents in 100g of foods 86
   origin of 22
   taste of 23
   which foods contain most 24
Systematically **34**

Tarnish **70**
TVP, classifying of 12

Vegetables
   making a coleslaw 6
   plant cells 8
   preparation of 3

Wall tiles, different uses of 55
Wash codes 74
Washing powders, investigation of 73
Washing-up 72
Watery foods
   water content of edible foods 9
   water content in 100g of foods 87
Whole grains, structure of 18